U0169640

高等学校大数据专业系列教材

大数据英语

实用教程

张强华　司爱侠　编著

西安电子科技大学出版社

内 容 简 介

本书主要介绍了大数据基础、大数据分析、结构化数据和半结构化数据以及非结构化数据、数据提取和转换及加载、数据库与数据库管理系统、数据仓库、数据集市、数据湖、云计算与大数据、数据处理、数据挖掘及其算法、大数据领域常用的编程语言、Hadoop 与 Spark、大数据可视化、大数据与人工智能、数据安全与数据隐私保护等方面的内容。全书共 10 个单元,各单元包含课文、单词、词组、缩写、难句解析、参考译文、习题、阅读材料等部分。

本书吸纳了作者近 20 年的 IT 行业英语翻译与图书编写经验,与课堂教学的各个环节紧密切合,支持备课、教学、复习及考试等各个教学环节,有配套的教学大纲、教学 PPT、习题参考答案及参考试题等。

本书既可作为高等院校大数据相关专业的专业英语教材,也可作为大数据相关从业人员的自学教材或者大数据相关培训班的教材。

图书在版编目(CIP)数据

大数据英语实用教程 / 张强华,司爱侠编著. —西安:西安电子科技大学出版社,2022.7
ISBN 978-7-5606-6494-1

Ⅰ. ①大… Ⅱ. ①张… ②司… Ⅲ. ①数据处理—英语—教材 Ⅳ. ①TP274

中国版本图书馆 CIP 数据核字(2022)第 086437 号

策 划 李惠萍
责任编辑 李惠萍
出版发行 西安电子科技大学出版社(西安市太白南路 2 号)
电 话 (029)88202421 88201467   邮 编 710071
网 址 www.xduph.com   电子邮箱 xdupfxb001@163.com
经 销 新华书店
印刷单位 陕西天意印务有限责任公司
版 次 2022 年 7 月第 1 版   2022 年 7 月第 1 次印刷
开 本 787 毫米×1092 毫米 1/16 印 张 15
字 数 353 千字
印 数 1~2000 册
定 价 36.00 元

ISBN 978-7-5606-6494-1 / TP

XDUP 6796001-1

如有印装问题可调换

前　言

　　大数据正在引领这个时代的技术发展，给社会带来了深刻的变革。近几年来，我国许多高校纷纷开设了大数据专业，以培养急需的专业人才。由于大数据行业的发展速度极快，从业人员必须掌握许多新技术、新方法、新软件，因此对专业英语要求较高。具备相关职业技能并精通专业外语的人员往往会赢得竞争优势，成为职场中不可或缺的核心人才与领军人物。本书就是为了满足当前大数据人才培养需要而编写的。本书的特点与优势如下：

　　(1) 选材全面，涉及大数据基础、大数据分析、结构化数据和半结构化数据以及非结构化数据、数据提取和转换及加载、数据库与数据库管理系统、数据仓库、数据集市、数据湖、云计算与大数据、数据处理、数据挖掘及其算法、大数据领域常用的编程语言、Hadoop 与 Spark、大数据可视化、大数据与人工智能、数据安全与数据隐私保护等各个方面。本书内容实用，覆盖面广泛，并与行业实际密切结合。编者对课文素材进行了严谨细致的编写，使其具有更强的实用性。

　　(2) 体例创新，适合教学。本书共 10 个单元，各单元主要包含以下部分：课文——给出选材广泛、风格多样、切合实际的两篇专业文章；单词——给出课文中出现的新词，读者由此可以积累大数据专业的基本词汇；词组——给出课文中的常用词组；缩略语——给出课文中出现的、业内人士必须掌握的缩略语；难句解析——讲解课文中的疑难语句，分析其语法结构，培养读者理解复杂语句的能力；参考译文——让读者对照理解课文提高读者的翻译能力；习题——包括针对课文的练习、词汇练习和翻译练习；阅读材料——进一步扩大读者的视野。

　　(3) 习题量适当，题型丰富，难易搭配，便于教师组织教学。

　　(4) 教学支持完善，为教师提供配套的教学大纲、教学 PPT、习题参考答案及参考试题等，教师可以在出版社网站免费下载。

(5) 为读者提供了配套的单词表、词组表和缩写表的电子版，扫描下面的二维码，即可下载。

(6) 作者有近 20 年 IT 行业英语图书的编写经验，这是作者编写的第三部大数据英语教材。在作者编写的英语书籍中，有三部国家级"十一五"规划教材，一部全国畅销书，一部获华东地区教材二等奖。这些图书编写经验有助于本书的完善与提升。

在使用本书的过程中，有任何问题都可以通过电子邮件与我们交流，我们一定会给予答复。邮件标题请注明姓名及"索取西安电子科技大学出版社大数据英语参考资料"字样。我们的 E-mail 地址为 zqh3882355@sina.com 和 zqh3882355@163.com。

由于作者水平有限，书中可能还存在不足之处，敬请读者批评指正。

作　者
2022 年 4 月

单词表

词组表

缩写表

目　　录

Unit 4

Unit 5

Unit 9

Unit 10

Unit 1

Bigdata **Text A**

What Is Big Data?

Big data is a term that describes datasets that are too large to be processed with the help of conventional tools and it is also sometimes used to call a field of study that concerns those datasets. In this passage, we will talk about the benefits of big data and how businesses can use it to succeed.

1. The Six Vs of Big Data

Big data is often described with the help of six Vs (see Figure 1-1). They allow us to understand the nature of big data better.

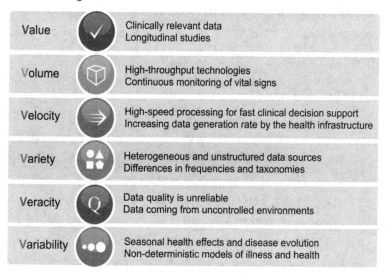

Figure 1-1   Six Vs of Big Data

1.1   Value

The meaning that you extract from data using special tools must bring real value by serving a specific goal, be it improving customer experience or increasing sales. For example, data that can be used to analyze consumer behavior is valuable for your company because you can use the research results to make individualized offers.

1.2 Volume

As its name suggests, big data is used to refer to enormous amounts of information. We are talking about not gigabytes but terabytes (1,099,511,627,776 bytes) and petabytes (1,125,899,906,842,624 bytes) of data.

1.3 Velocity

Velocity means that big data should be processed fast, in a stream-like manner because it just keeps coming. For example, a single Jet engine generates more than 10 terabytes of data in 30 minutes of flight time. Now imagine how much data you would have to collect to research one small airline company. Data never stops growing, and every new day you have more information to process than yesterday. This is why working with big data is so complicated.

1.4 Variety

Big data is usually not homogeneous. For example, the data of an enterprise consists of emails, documentation, images, photos, transaction records, etc. In order to derive any insights from this data, you need to classify and organize it first.

1.5 Veracity

Veracity describes whether the data can be trusted. Hygiene of data in analytics is important because otherwise, you cannot guarantee the accuracy of your results.

1.6 Variability

Variability describes how fast and to what extent data under investigation is changing. This parameter is important because even small deviations in data can affect the results. If the variability is high, you will have to constantly check whether your conclusions are still valid.

2. Why Big Data Is Important?

With the development and increase of apps and social media as well as more and more people and businesses moving online, there's been a huge increase in data. If we look at social media platforms, we'll find they interest and attract over servery million users daily, scaling up data more than ever before. The next question is how exactly this huge amount of data is handled and how it is processed and stored. This is where big data comes into play.

Big data analytics has revolutionized the field of IT, enhancing and adding added advantage to organizations. It involves the use of analytics, such as machine learning, mining, statistics and more. Big data can help organizations and teams to perform multiple operations on a single platform, store Tbs of data, pre-process it , analyze all the data, irrespective of the size and type, and visualize it, too.

3. How Big Data Works?

Big data can be categorized as unstructured or structured. Structured data consists of information already managed by the organization in databases and spreadsheets; it is frequently numeric in nature. Unstructured data is information that is unorganized and does not fall into a predetermined model or format. It includes data gathered from social media sources which help institutions gather information on customer needs.

Big data can be collected from publicly shared comments on social networks and websites, voluntarily gathered from personal electronics and apps, through questionnaires, product purchases and electronic check-ins. The presence of sensors and other inputs in smart devices allows data to be gathered across a broad spectrum of situations and circumstances.

Big data is most often stored in computer databases and is analyzed using software specifically designed to handle large, complex data sets. Many software as a service (SaaS) companies specialize in managing this type of complex data.

4. Benefits of Big Data

Big data analytics allows you to look deeper.

Very often, important decisions in production or management are made based on personal opinions or unconfirmed facts. By analyzing data, you will get objective insights into how things really are.

For example, big data analytics is now more and more widely used for rating employees for HR purposes. Imagining that you want to make one of the managers a vice-president, but don't know which one to choose, data analytics algorithms can analyze hundreds of parameters, such as when they start and finish their workday, what apps they use during the day, etc., to help you make decision.

Big data analytics helps you to optimize your resources, perform better risk management, and be data-driven when setting business goals.

5. Big Data Challenges

Understanding big data is challenging. It seems that its possibilities are limitless, and, indeed, we have many great solutions that rely heavily on big data. A few of those are recommender systems on Netflix, YouTube, or Spotify that all of us know and love. Often, we may not like their recommendations, but, in many cases, they are valuable.

6. How to Use Big Data?

If you want to benefit from the usage of big data, follow these steps.

6.1　Set a big data strategy

First, you need to set up a strategy. That means you need to identify what you want to achieve, for example, provide a better customer experience, improve sales, or improve your marketing strategy by learning more about the behavioral patterns of your clients. The goal defines the tools and data you will use for your research.

Let's say you want to study brand awareness of your company. For that, you will conduct social analytics and process raw unstructured data from various social media and review websites like Facebook, Twitter and Instagram. This type of analytics allows assessing brand awareness, measuring engagement, and seeing how word-of-mouth works for you.

In order to make the most out of your research, it is a good idea to assess the state of your company before analyzing. For example, you can collect the assumptions about your marketing strategy in social media and stats from different tools so that you can compare them with the results of your data-driven research and make conclusions.

6.2 Access and analyze the data

Once you have identified your goals and data sources, it is time to collect and analyze data. Very often, you have to preprocess it first so that machine learning algorithms could understand it.

By applying textual analysis, cluster analysis, predictive analytics and other methods of data mining, you can extract valuable insights from the data.

6.3 Make data-driven decisions

Use what you have learned about your business or another area of study in practice. The data-driven approach is already adopted by many countries all around the world. Insights taken from data allow you not to miss any important opportunities and manage your resources with maximum efficiency.

New Words

dataset	['deɪtəset]	n. 数据集
conventional	[kən'venʃənl]	adj. 通常的；传统的
nature	['neɪtʃə]	n. 天性
volume	['vɒljuːm]	n. 大量
information	[ˌɪnfə'meɪʃn]	n. 信息，消息
gigabyte	['gɪgəbaɪt]	n. 吉字节，千兆字节，十亿字节
terabyte	['terəbaɪt]	n. 太字节
petabyte	['petəbaɪt]	n. 拍字节
velocity	[və'lɒsəti]	n. 高速，快速；速率，速度
engine	['endʒɪn]	n. 引擎
generate	['dʒenəreɪt]	v. 产生；形成，造成
imagine	[ɪ'mædʒɪn]	v. 想象；猜测
complicate	['kɒmplɪkeɪt]	v. 使复杂化
variety	[və'raɪəti]	n. 多样化，多种多样

homogeneous	[ˌhɒməˈdʒiːnɪəs]	adj. 同性质的，同类的
enterprise	[ˈentəpraɪz]	n. 公司，企(事)业单位
documentation	[ˌdɒkjʊmenˈteɪʃn]	n. 文档
derive	[dɪˈraɪv]	v. (使)起源于，来自；获得
insight	[ˈɪnsaɪt]	n. 洞察力；领悟；见解
classify	[ˈklæsɪfaɪ]	v. 将……分类
improve	[ɪmˈpruːv]	v. 改进，提高
analyze	[ˈænəlaɪz]	v. 分析
individualize	[ˌɪndɪˈvɪdʒʊəlaɪz]	v. 赋予个性，个别地加以考虑
veracity	[vəˈræsəti]	n. 真实性
variability	[ˌveərɪəˈbɪləti]	n. 变化性，易变性，可变性
describe	[dɪˈskraɪb]	v. 描述，把……称为
parameter	[pəˈræmɪtə]	n. 参数；限制因素；决定因素
affect	[əˈfekt]	v. 影响
constantly	[ˈkɒnstəntli]	adv. 不断地，时常地；始终，一直
check	[tʃek]	v. 检查；查看；核实
platform	[ˈplætfɔːm]	n. 平台；论坛
attract	[əˈtrækt]	v. 吸引
handle	[ˈhændl]	v. 处理，应付
process	[ˈprəʊses]	v. 加工；处理
		n. 过程
store	[stɔː]	v. 存储，保存
statistics	[stəˈtɪstɪks]	n. 统计学
operation	[ˌɒpəˈreɪʃn]	n. 活动；操作
irrespective	[ɪrɪˈspektɪv]	adj. 不考虑的，不顾的；无关的
visualize	[ˈvɪʒʊəlaɪz]	v. 可视化，使可见
structured	[ˈstrʌktʃəd]	adj. 结构化的
unstructured	[ʌnˈstrʌktʃəd]	adj. 非结构化的
spreadsheet	[ˈspredʃiːt]	n. 电子表格
unorganized	[ʌnˈɔːgənaɪzd]	adj. 无组织的
predetermine	[ˌpriːdɪˈtɜːmɪn]	v. 事先安排，预先设定
gather	[ˈgæðə]	v. 收集，采集
institution	[ˌɪnstɪˈtjuːʃn]	n. 机构
collect	[kəˈlekt]	v. 收集；获得
website	[ˈwebsaɪt]	n. 网站
voluntarily	[ˈvɒləntərɪli]	adv. 志愿地；自动地，自发地
sensor	[ˈsensə]	n. 传感器
input	[ˈɪnpʊt]	n.&v. 输入
spectrum	[ˈspektrəm]	n. 光谱，波谱；范围

software	['sɒftweə]	n. (计算机)软件
unconfirmed	[ˌʌnkən'fɜ:md]	adj. 未经证实的，未予确认的
objective	[əb'dʒektɪv]	adj. 客观的；基于事实的
		n. 目标
algorithm	['ælgərɪðəm]	n. 算法
decision	[dɪ'sɪʒn]	n. 决定，决策
optimize	['ɒptɪmaɪz]	v. 使最优化
data-driven	['deɪtə 'drɪvn]	adj. 数据驱动的
possibility	[ˌpɒsə'bɪləti]	n. 可能性
limitless	['lɪmɪtləs]	adj. 无限制的；无界限的
recommendation	[ˌrekəmen'deɪʃn]	n. 推荐，建议
pattern	['pætn]	n. 模式；范例
access	['ækses]	v. 访问，存取
preprocess	[pri:'prəuses]	v. 预加工，预处理
cluster	['klʌstə]	n. 丛；簇
predictive	[prɪ'dɪktɪv]	adj. 预测的，预言性的
adopt	[ə'dɒpt]	v. 采用
maximum	['mæksɪməm]	adj. 最大的
		adv. 最多地
		n. 最大量

Phrases

big data	大数据
flight time	飞行时间
airline company	航空公司
consist of ...	由……组成
transaction record	事务记录，交易记录
in order to ...	为了……
social media	社交媒体
scale up	按比例加大
machine learning	机器学习
structured data	结构化数据
unstructured data	非结构化数据
fall into	分成
smart device	智能设备
base on	基于
recommender system	推荐系统

set up	建立，创建；安排
behavioral pattern	行为模式
brand awareness	商标意识，品牌意识
data source	数据源
textual analysis	文本分析
data mining	数据挖掘
in practice	在实践中；在练习中

Abbreviations

app (application)	应用程序
IT (Information Technology)	信息技术
SaaS (Software as a Service)	软件即服务
HR (Human Resource)	人力资源

Analysis of Difficult Sentences

[1] Big data is a term that describes datasets that are too large to be processed with the help of conventional tools and it is also sometimes used to call a field of study that concerns those datasets.

本句中，and 连接了两个并列句。that describes datasets that are too large to be processed with the help of conventional tools 是定语从句，修饰和限定 a term。在该从句中，that are too large to be processed with the help of conventional tools 是定语从句，修饰和限定 datasets。too... to... 的意思是"太……以至于不能……"。that concerns those datasets 是一个定语从句，修饰和限定 a field of study。

[2] The meaning that you extract from data using special tools must bring real value by serving a specific goal, be it improving customer experience or increasing sales.

本句中，that you extract from data using special tools 是一个同位语从句，对 The meaning 进行补充说明。by serving a specific goal 是一个介词短语作方式状语，be it improving customer experience or increasing sales 是一个让步状语从句，它们都修饰谓语 must bring real value。

[3] For example, data that can be used to analyze consumer behavior is valuable for your company because you can use the research results to make individualized offers.

本句中，that can be used to analyze consumer behavior 是一个定语从句，修饰和限定 data。because you can use the research results to make individualized offers 是一个原因状语从句，修饰谓语 is valuable。

[4] For example, you can collect the assumptions about your marketing strategy in social

media and stats from different tools so that you can compare them with the results of your data-driven research and make conclusions.

本句中，so that you can compare them with the results of your data-driven research and make conclusions 是一个目的状语从句，修饰谓语 collect。

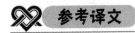 参考译文

什么是大数据？

大数据是一个术语，用于描述过大而无法借助传统工具进行处理的数据集，有时也被用来称呼与这些数据集有关的研究领域。在这篇文章中，我们将讨论大数据的好处以及企业如何利用大数据取得成功。

1. 大数据的六个 V

大数据通常借助六个 V 来描述(见图 1-1)。它们使我们能够更好地了解大数据的本质。

图 1-1 大数据的六个 V

1.1 价值

使用特殊工具从数据中提取的含义必须通过服务于特定目标来带来真正的价值，无论是为了改善客户体验还是增加销售额。例如，可用于分析消费者行为的数据对你的公司很有价值，因为可以使用研究结果进行个性化报价。

1.2 大量

顾名思义，大数据是指海量的信息。我们谈论的不是千兆字节，而是太字节

(1 099 511 627 776 字节)和拍字节(1 125 899 906 842 624 字节)的数据。

1.3 高速

高速意味着大数据应该以类似流的方式快速处理，因为数据不断涌现。例如，单个 Jet 引擎在 30 分钟的飞行时间内生成超过 10 TB 的数据。想象一下，研究一家小型航空公司需要收集多少数据。数据永远不会停止增长，每一天都有比昨天更多的信息需要处理。这就是处理大数据如此复杂的原因。

1.4 多样性

大数据通常不是同质的。例如，企业的数据包括其电子邮件、文档、图像、照片、交易记录等。为了从这些数据中获得任何见解，首先需要对其进行分类和组织。

1.5 真实性

真实性描述数据是否可信。分析中数据的清洁很重要，因为如果数据不清洁，你将无法保证结果的准确性。

1.6 易变性

易变性描述了调查数据变化的速度和程度。此参数很重要，因为即使数据中的微小偏差也会影响结果。如果易变性很高，就必须不断检查结论是否仍然有效。

2. 为什么大数据如此重要？

随着应用程序和社交媒体的发展与增加，以及越来越多的人和企业上网，数据出现了巨大的增长。如果查看社交媒体平台会发现它们每天都吸引着超过百万的用户，比以往任何时候都更能扩展数据。下一个问题是如何处理如此庞大的数据量以及如何存储和加工这些数据。这就是大数据发挥作用的地方。

大数据分析彻底改变了 IT 领域，加强并增加了组织的优势。大数据分析涉及分析的使用，例如机器学习、挖掘、统计等。大数据可以帮助组织和团队在单个平台上执行多项操作、存储太字节的数据、对其进行预处理、分析所有数据而不考虑其大小和类别并将其可视化。

3. 大数据如何运作？

大数据可分为非结构化或结构化数据。结构化数据由组织已经在数据库和电子表格中管理好的信息组成，它通常是数字性质的。非结构化数据是无组织的且不属于预定模型或格式。它包括从社交媒体中收集的数据，这些数据可帮助机构收集有关客户需求的信息。

大数据可以从社交网络和网站上公开共享的评论中收集，从个人电子产品和应用程序中自愿收集，也可以通过问卷调查、产品购买和电子签到的方式收集。智能设备中传感器和其他输入的存在允许广泛收集数据。

大数据通常存储在计算机数据库中，并使用专门用于处理大型复杂数据集的软件进行分析。许多软件即服务(SaaS)公司专门管理此类复杂数据。

4．大数据的好处

大数据分析使你可以更深入地了解事物。

很多时候，生产或管理方面的重要决策都是基于个人观点或未经证实的事实做出的。通过分析数据可以客观地了解事物的真实情况。

例如，大数据分析现在越来越广泛地用于为人力资源目的对员工进行评级。想象一下，你想让某一位经理成为副总裁，但不知道该选择谁。数据分析算法可以分析数百个参数，例如他们开始和结束工作日的时间、他们在一天中使用的应用程序等，以帮助你做出决定。

大数据分析可以帮助你优化资源、进行更好的风险管理，并在设定业务目标时以数据为导向。

5．大数据挑战

理解大数据具有挑战性。似乎它的可能性是无限的，事实上，我们有许多非常依赖大数据的优秀解决方案。其中一些是我们都知道和喜爱的 Netflix、YouTube 或 Spotify 上的推荐系统。通常，我们可能不喜欢这些建议，但在许多情况下，它们是有价值的。

6．如何使用大数据？

如果你想从大数据的使用中受益，可按照以下步骤进行操作。

6.1 制定大数据策略

首先，需要制定策略。这意味着需要确定想要实现的目标，例如，提供更好的客户体验、提高销售或通过更深入了解客户的行为模式改进营销策略。目标决定了你将用于研究的工具和数据。

假设你想研究公司的品牌知名度。为此，你将进行社交分析并处理来自各种社交媒体和评论网站(如 Facebook、Twitter 和 Instagram)的原始非结构化数据。这种类型的分析可以评估品牌知名度、衡量参与度并了解口碑如何发挥作用。

为了充分利用你的研究，最好在分析之前评估你公司的状况。例如，你可以在社交媒体中收集有关营销策略的假设，用不同的工具收集统计数据，以便可以将它们与数据驱动的研究结果进行比较并得出结论。

6.2 访问和分析数据

一旦确定了目标和数据源，就该收集和分析数据了。通常，必须先对其进行预处理，以便机器学习算法能够理解。

通过应用文本分析、聚类分析、预测分析和其他数据挖掘方法，可以从数据中提取有价值的见解。

6.3 做出数据驱动的决策

在实践中运用你所学到的关于业务或其他学习领域的知识。数据驱动的方法已经被世界上许多国家采用。从数据中获取的见解让你不会错过重要机会并以最高效率管理资源。

Bigdata **Text B**

Structured Data, Unstructured Data and Semi-structured Data

All data is not created equal. Some data is structured, but most of it is unstructured. Structured and unstructured data is sourced, collected and scaled in different ways, and each one resides in a different type of database.

1. What Is Structured Data?

Structured data—typically categorized as quantitative data—is highly organized and easily decipherable by machine learning algorithms. Examples of structured data include dates, names, addresses, credit card numbers, etc (see Figure 1-2). Developed by IBM in 1974, structured query language (SQL) is the programming language used to manage structured data. By using a relational database, business users can quickly input, search and manipulate structured data.

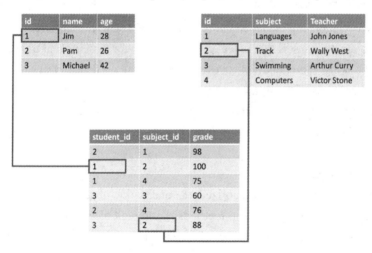

Figure 1-2   Structured Data Example

1.1   Pros and cons of structured data

Pros:

• Easily used by machine learning (ML) algorithms: The specific and organized architecture of structured data eases manipulation and querying of ML data.

• Easily used by business users: Structured data does not require an in-depth understanding of different types of data and how they function. With a basic understanding of the topic relative to the data, users can easily access and interpret the data.

• Accessible by more tools: Since structured data predates unstructured data, there are more tools available for using and analyzing structured data.

Cons:

• Limited usage: Data with a predefined structure can only be used for its intended purpose, which limits its flexibility and usability.

• Limited storage options: Structured data is generally stored in data storage systems with rigid schemas. Therefore, changes in data requirements necessitate an update of all structured data, which leads to a massive expenditure of time and resources.

1.2 Structured data tools

• OLAP: Performs high-speed, multidimensional data analysis from unified, centralized data stores.

• SQLite: Implements a self-contained, serverless, zero-configuration transactional relational database engine.

• MySQL: Embeds data into mass-deployed software, particularly mission-critical, heavy-load production systems.

• PostgreSQL: Supports SQL and JSON querying as well as high-tier programming languages (C/C+, Java, Python, etc.).

1.3 Use cases for structured data

• Customer relationship management (CRM): CRM software runs structured data through analytical tools to create datasets that reveal customer behavior patterns and trends.

• Online booking: Hotel and ticket reservation data (e.g., dates, prices, destinations, etc.) fits the "rows and columns" format indicative of the predefined data model.

• Accounting: Accounting firms or departments use structured data to process and record financial transactions.

2. What Is Unstructured Data?

Unstructured data, typically categorized as qualitative data, cannot be processed and analyzed via conventional data tools and methods. Examples of unstructured data include text, mobile activity, social media posts, Internet of Things (IoT) sensor data, etc. The importance of unstructured data is increasing rapidly. It is predicted that unstructured data accounts for more than 80% of all enterprise data, and 95% of enterprises give priority to unstructured data management. Since unstructured data does not have a predefined data model, it is best managed in non-relational databases. Another way to manage unstructured data is to use data lakes to preserve it in raw form.

2.1 Pros and cons of unstructured data

Pros:

• Native format: Unstructured data, stored in its native format, remains undefined until needed. Its adaptability increases file formats in the database, which widens the data pool and enables data scientists to prepare and analyze only the data they need.

• Fast accumulation rates: Since there is no need to predefine the data, it can be collected quickly and easily.

• Data lake storage: Allows for massive storage and pay-as-you-use pricing, which cuts costs and eases scalability.

Cons:

• Required expertise: Due to its undefined/non-formatted nature, data science expertise is required to prepare and analyze unstructured data. This is beneficial to data analysts but alienates unspecialized business users who may not fully understand specialized data topics or how to utilize their data.

• Specialized tools: Specialized tools are required to manipulate unstructured data, which limits product choices for data managers.

2.2   Unstructured data tools

• MongoDB: Uses flexible documents to process data for cross-platform applications and services.

• DynamoDB: Delivers single-digit millisecond performance at any scale via built-in security, in-memory caching and backup and restore.

• Hadoop: Provides distributed processing of large data sets using simple programming models and no formatting requirements.

• Azure: Enables agile cloud computing for creating and managing apps through Microsoft's data centers.

2.3   Use cases for unstructured data

• Data mining: Enables businesses to use unstructured data to identify consumer behavior, product sentiment, and purchasing patterns to better accommodate their customer base.

• Predictive data analytics: Alerts businesses of important activity ahead of time so they can properly plan and accordingly adjust to significant market shifts.

• Chatbots: Performs text analysis to route customer questions to the appropriate answer sources.

3.   What Are the Key Differences Between Structured and Unstructured Data?

While structured (quantitative) data gives a "birds-eye view" of customers, unstructured (qualitative) data provides a deeper understanding of customer behavior and intent. Let's explore some of the key areas of difference and their implications. (see Figure 1-3)

• Sources: Structured data is sourced from GPS sensors, online forms, network logs, web server logs, OLTP systems, etc., whereas unstructured data sources include email messages, word-processing documents, PDF files, etc.

• Forms: Structured data consists of numbers and values, whereas unstructured data consists of sensors, text files, audio and video files, etc.

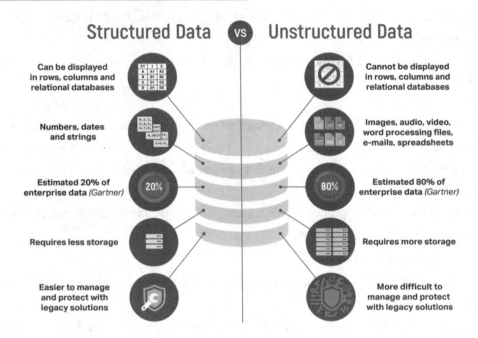

Figure 1-3　Structured Data vs. Unstructured Data

• Models: Structured data has a predefined data model and is formatted to a set data structure before being placed in data storage (e.g., schema on write), whereas unstructured data is stored in its native format and not processed until it is used (e.g., schema on read).

• Storage: Structured data is stored in tabular formats (e.g., excel sheets or SQL databases) that require less storage space. It can be stored in data warehouses, which makes it highly scalable. Unstructured data, on the other hand, is stored as media files or No SQL databases, which require more space. It can be stored in data lakes which makes it difficult to scale.

• Uses: Structured data is used in machine learning and drives its algorithms, whereas unstructured data is used in natural language processing (NLP) and text mining.

4.　What Is Semi-structured Data?

Semi-structured data (e.g., JSON, CSV, XML) is the "bridge" between structured and unstructured data. It does not have a predefined data model and is more complex than structured data, yet easier to store than unstructured data.

Semi-structured data uses "metadata" (e.g., tags and semantic markers) to identify specific data characteristics and scale data into records and preset fields. Metadata ultimately enables semi-structured data to be better cataloged, searched and analyzed than unstructured data.

Example of metadata usage: An online article displays a headline, a snippet, a featured image, image alt text, etc., which helps differentiate one piece of web content from similar pieces.

Example of semi-structured data vs. structured data: A tab-delimited file containing customer data versus a database containing CRM tables.

Example of semi-structured data vs. unstructured data: A tab-delimited file versus a list of comments from a customer's Instagram.

5. The Future of Data

Recent developments in artificial intelligence (AI) and machine learning are driving the future wave of data, which is enhancing business intelligence and advancing industrial innovation. In particular, the data formats and models covered in this article are helping business users to do the following:

• Analyze digital communications for compliance: Pattern recognition and email threading analysis software can search email and chat data for potential noncompliance.

• Track high-volume customer conversations in social media: Text analytics and sentiment analysis enables monitoring of marketing campaign results and identifying online threats.

• Gain new marketing intelligence: ML analytics tools can quickly cover massive amounts of data to help businesses analyze customer behavior.

Furthermore, smart and efficient usage of data formats and models can help you with the following:

• Understand customer needs at a deeper level to better serve them;

• Create more focused and targeted marketing campaigns;

• Track current metrics and create new ones;

• Create better product opportunities and offerings;

• Reduce operational costs.

New Words

semi-structured	['semɪ 'strʌktʃəd]	adj. 半结构化的
scale	[skeɪl]	v. 改变……的大小
		n. 规模，等级
categorize	['kætəgəraɪz]	v. 把……归类，把……分门别类
decipherable	[dɪ'saɪfərəbl]	adj. 可判读的；可翻译的
search	[sɜːtʃ]	n.&v. 搜索，检索；查找
architecture	['ɑːkɪtektʃə]	n. 体系结构；架构
manipulation	[məˌnɪpjʊ'leɪʃn]	n. (熟练的)操作；操纵；控制
function	['fʌŋkʃn]	n. 功能；函数
predefined	[priːdɪ'faɪnd]	adj. 预定义的
limit	['lɪmɪt]	v. 限制；限量
		n. 极限；限制；界限
flexibility	[ˌfleksə'bɪləti]	n. 灵活性；弹性

usability	[ˌjuːzə'bɪlɪti]	n. 可用性；适用性
storage	['stɔːrɪdʒ]	n. (计算机的)存储
option	['ɒpʃn]	n. 选择；选择权；可选择的事物
rigid	['rɪdʒɪd]	adj. 严格的
schema	['skiːmə]	n. 概要，计划
necessitate	[nə'sesɪteɪt]	v. 使……成为必要，需要；强迫
update	[ˌʌp'deɪt]	v. 更新，升级
massive	['mæsɪv]	adj. 巨大的；大而重的；强烈的
expenditure	[ɪk'spendɪtʃə]	n. 花费；消耗
perform	[pə'fɔːm]	v. 执行；起……作用
multidimensional	[ˌmʌltɪdɪ'menʃənl]	adj. 多维的，多面的
unify	['juːnɪfaɪ]	v. 统一；使联合，使一致
self-contained	[self kən'teɪnd]	adj. 独立的；自给自足的
serverless	['sɜːvələs]	adj. 无服务器的
embed	[ɪm'bed]	v. (使)嵌入，融入
heavy-load	['hevɪ ləʊd]	n. 重载，高负载
priority	[praɪ'ɒrəti]	n. 优先，优先权；重点
native	['neɪtɪv]	adj. 天生的；本地的
adaptability	[əˌdæptə'bɪlɪti]	n. 适应性；合用性
file	[faɪl]	n. 文件
accumulation	[əˌkjuːmjə'leɪʃn]	n. 堆积(物)；积累
scalability	[skeɪlə'bɪlɪti]	n. 可扩展性
alienate	['eɪlɪəneɪt]	v. 使不友好；使疏远
unspecialized	['ʌn'speʃəlaɪzd]	adj. 非专业化的
utilize	['juːtəlaɪz]	v. 利用，使用
built-in	[ˌbɪlt'ɪn]	adj. 嵌入的；内置的；固有的
cache	[kæʃ]	n. 高速缓冲储存区
backup	['bækʌp]	n. 备份，备用品
restore	[rɪ'stɔː]	v. 恢复；修复
distributed	[dɪs'trɪbjuːtɪd]	adj. 分布式的
agile	['ædʒaɪl]	adj. 敏捷的，灵巧的
alert	[ə'lɜːt]	v. 向……报警；使警觉，使戒备
		n. 警惕，戒备；警报
		adj. 警觉的，警惕的
adjust	[ə'dʒʌst]	v. 调整，调节
significant	[sɪg'nɪfɪkənt]	adj. 重要的；显著的
network	['netwɜːk]	n. 网络

form	[fɔːm]	v. 形成，构成
		n. 类型；表格
tabular	['tæbjələ]	adj. 表格的
metadata	['metədeɪtə]	n. 元数据
semantic	[sɪ'mæntɪk]	adj. 语义的，语义学的
characteristic	[ˌkærəktə'rɪstɪk]	n. 特色，特点
preset	[ˌpriː'set]	v. 预先布置；事先安排
snippet	['snɪpɪt]	n. 小片，片段
compliance	[kəm'plaɪəns]	n. 合规性
noncompliance	[ˌnɒnkəm'plaɪəns]	n. 不合规性
monitor	['mɒnɪtə]	v. 监控；监听
		n. 显示屏；监测仪器
smart	[smɑːt]	adj. 智能的；敏捷的

Phrases

semi-structured data	半结构化数据
reside in …	驻留在……
credit card	信用卡
programming language	编程语言，程序设计语言
be used for ...	用来做……；用于……
data storage system	数据存储系统
data requirement	数据要求，数据需求
mass-deployed software	大规模部署软件
data model	数据模型
non-relational database	非关系数据库
raw form	原始形式
data pool	数据池
business user	商业用户，业务用户
cloud computing	云计算
data center	数据中心
purchasing pattern	购买模式
customer base	客户群
market shift	市场转变
birds-eye view	鸟瞰图
text file	文本文件

schema on write	写时模式
schema on read	读时模式
media file	媒体文件
image alt text	图片替代文本
business intelligence	商业智能
pattern recognition	模式识别
email threading analysis	电子邮件线程分析
sentiment analysis	情感分析
operational cost	经营成本，运营成本

Abbreviations

ML (Machine Learning)	机器学习
OLTP (On-Line Transaction Processing)	联机事务处理
JSON (JavaScript Object Notation)	JS 对象简谱
CRM (Customer Relationship Management)	客户关系管理
IoT (Internet of Things)	物联网
GPS (Global Position System)	全球定位系统
PDF(Portable Document Format)	便携文件格式
NLP (Natural Language Processing)	自然语言处理
CSV (Comma Separated Value)	逗号分隔值
XML (eXtensible Markup Language)	可扩展标记语言
AI (Artificial Intelligence)	人工智能

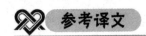 参考译文

结构化数据、非结构化数据和半结构化数据

所有数据并不都是相同的。一些数据是结构化的，但大部分是非结构化的。结构化和非结构化数据的来源、收集和缩放方式各不相同，并且每一种数据都驻留在不同类型的数据库中。

1. 什么是结构化数据？

结构化数据(通常被归类为定量数据)是高度组织化的，并且可以通过机器学习算法轻松解读。结构化数据的示例包括日期、姓名、地址、信用卡号等(见图 1-2)。IBM 于 1974

年开发的结构化查询语言(SQL)是用于管理结构化数据的编程语言。通过使用关系数据库,业务用户可以快速输入、搜索和操作结构化数据。

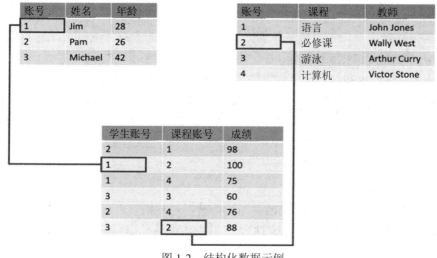

图 1-2   结构化数据示例

1.1   结构化数据的优缺点

结构化数据的优点如下:

• 容易被机器学习(ML)算法所使用:结构化数据的特点及有组织的架构简化了 ML 数据的操作和查询。

• 容易被业务用户所使用:结构化数据不需要深入了解不同类型的数据及其运作方式。用户只要基本了解与数据相关的主题,就可以轻松访问和解释数据。

• 可通过更多工具访问:由于结构化数据早于非结构化数据,因此有更多工具可用于使用和分析结构化数据。

结构化数据的缺点如下:

• 使用受限:具有预定义结构的数据只能用于预期目的,这限制了灵活性和可用性。

• 有限的存储选择:结构化数据通常存储在具有严格架构的数据存储系统中。因此,数据需求的变了,就需要更新所有结构化数据,这会造成大量的时间和资源消耗。

1.2   结构化数据工具

• OLAP:从统一、集中的数据存储中执行高速、多维数据分析。

• SQLite:实现独立的、无服务器的、零配置的事务关系数据库引擎。

• MySQL:将数据嵌入到大规模部署的软件中,尤其是任务关键型、高负载的生产系统。

• PostgreSQL:支持 SQL 和 JSON 查询以及高级编程语言(C/C+、Java、Python 等)。

1.3   结构化数据的用例

• 客户关系管理(CRM):CRM 软件通过分析工具运行结构化数据,以创建揭示客户行为模式和趋势的数据集。

• 在线预订:酒店和机票预订数据(例如日期、价格、目的地等)符合表示预定义数据模型的"行和列"格式。

- 会计：会计师事务所或部门使用结构化数据来处理和记录财务交易。

2. 什么是非结构化数据？

非结构化数据通常归类为定性数据，无法通过传统的数据工具和方法进行处理和分析。非结构化数据的示例包括文本、移动活动、社交媒体帖子、物联网 (IoT) 传感器数据等。非结构化数据的重要性正在迅速增加。据预测，非结构化数据占企业所有数据的 80% 以上，95% 的企业优先使用非结构化数据管理。由于非结构化数据没有预定义的数据模型，因此最好在非关系数据库中进行管理。管理非结构化数据的另一种方法是使用数据湖以原始形式保存它。

2.1 非结构化数据的优缺点

非结构化数据的优点如下：

- 原生格式：以原生格式存储的非结构化数据在被需要前保持未定义状态。它的适应性增加了数据库中的文件格式，从而扩大了数据池并使数据科学家能够仅准备和分析他们需要的数据。
- 快速积累率：由于无须预先定义数据，因此可以快速轻松地收集数据。
- 数据湖存储：支持海量存储和按使用付费的定价，从而降低成本并简化可扩展性。

非结构化数据的缺点如下：

- 需要专业知识：由于其未定义/非格式化的性质，需要数据科学专业知识来准备和分析非结构化数据。这对数据分析师是有益的，但会疏远非专业化的业务用户，他们可能不完全了解专业化的数据主题或如何利用他们的数据。
- 专业工具：需要专业工具来处理非结构化数据，这限制了数据管理者的产品选择。

2.2 非结构化数据工具

- MongoDB：使用灵活的文档来处理跨平台应用程序和服务的数据。
- DynamoDB：通过内置安全性、内存中缓存以及备份和恢复，在任何规模下提供几毫秒响应的性能。
- Hadoop：使用简单的编程模型和无格式要求提供大型数据集的分布式处理。
- Azure：支持敏捷的云计算，通过微软的数据中心创建和管理应用程序。

2.3 非结构化数据的用例

- 数据挖掘：使企业能够使用非结构化数据来识别消费者行为、产品意见和购买模式，以更好地适应其客户群。
- 预测性数据分析：提前提醒业务部门重要活动，以便正确规划并相应调整以适应重大的市场变化。
- 聊天机器人：执行文本分析，将客户问题连接到适当的答案。

3. 结构化和非结构化数据的主要区别是什么？

虽然结构化(定量)数据提供了客户的"鸟瞰图"，但非结构化(定性)数据提供了对客户行为和意图的更深入了解。下面探讨一些关键的差异领域及其影响(见图1-3)。

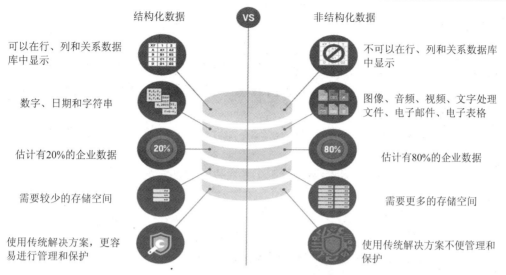

图 1-3　结构化数据与非结构化数据对照

· 来源：结构化数据来自 GPS 传感器、在线表格、网络日志、Web 服务器日志、OLTP 系统等，而非结构化数据来源包括电子邮件消息、文字处理文档、PDF 文件等。

· 构成：结构化数据由数字和值组成，而非结构化数据由传感器、文本文件、音频和视频文件等组成。

· 模型：结构化数据具有预定义的数据模型，并在进行数据存储(如写时模式)之前格式化为设定的数据结构，而非结构化数据以其原生格式存储直到使用时才进行处理(如读时模式)。

· 存储：结构化数据以需要较少存储空间的表格格式(如Excel 表格或SQL 数据库)存储。它可以存储在数据仓库中，这使其具有高度的可扩展性。另一方面，非结构化数据存储为媒体文件或 NoSQL 数据库，这需要更多空间。它可以存储在数据湖中，这使得它难以扩展。

· 用途：结构化数据用于机器学习并驱动其算法，而非结构化数据用于自然语言处理(NLP)和文本挖掘。

4．什么是半结构化数据？

半结构化数据(如 JSON、CSV、XML)是结构化和非结构化数据之间的"桥梁"。它没有预定义的数据模型，比结构化数据更复杂，但比非结构化数据更容易存储。

半结构化数据使用"元数据"(如标签和语义标记)来识别特定的数据特征并将数据缩放到记录和预设字段中。元数据最终使半结构化数据能够比非结构化数据更好地进行编目、搜索和分析。

元数据使用示例：在线文章显示标题、摘要、特色图片、图片替代文本等，这有助于将网络内容与类似部分区分开来。

半结构化数据与结构化数据的对比示例：前者包含客户数据的制表符分隔文件，后者包含 CRM 表的数据库。

半结构化数据与非结构化数据的对比示例：前者是制表符分隔的文件，后者是来自客户 Instagram 的评论列表。

5. 数据的未来

人工智能(AI)和机器学习的最新发展正在推动未来的数据浪潮，这将增强商业智能并推动工业创新。尤其本文中涵盖的数据格式和模型正在帮助业务用户执行以下操作：

- 分析数字通信的合规性：模式识别和电子邮件线程分析软件可以搜索电子邮件和聊天数据以查找潜在的不合规情况。

- 跟踪社交媒体中的大量客户对话：文本分析和情绪分析，可以监控营销活动结果并识别在线威胁。

- 获得新的营销情报：机器学习分析工具可以快速涵盖海量数据，帮助企业分析客户行为。

此外，智能高效地使用数据格式和模型可以帮助你完成以下工作：

- 更深层次地了解客户需求以更好地为他们服务；
- 创建更有针对性和目标性的营销活动；
- 跟踪当前指标并创建新指标；
- 创造更好的产品机会和产品；
- 降低运营成本。

𝓑𝒾𝓰𝒹𝒶𝓉𝒶 **Exercises**

〖Ex.1〗 根据 Text A 回答以下问题。

1. What are the six vs. that are usd to describe big data?
2. What does velocity mean?
3. What does the data of an enterprise consist of?
4. What will we find if we look at social media platforms?
5. What can big data be categorized as? What does structured data consist of?
6. What is unstructured data? What does it include?
7. What are important decisions in production or management made based on very often?
8. What does big data analytics help you to do?
9. What will your goal define?
10. What do insights taken from data allow you?

〖Ex. 2〗 根据 Text B 回答以下问题。

1. What do examples of structured data include?
2. What are the pros of structured data?
3. What are the cons of structured data?
4. What do examples of unstructured data include?
5. What are the pros of unstructured data?
6. What are the cons of unstructured data?

7. What are use cases for unstructured data?

8. Where is structured data sourced from?

9. What is semi-structured data?

10. What are the data formats and models covered in this article helping business users to do in particular?

〖Ex. 3〗 把下列词组翻译成中文。

1.	big data	1. _____
2.	base on	2. _____
3.	structured data	3. _____
4.	textual analysis	4. _____
5.	recommender system	5. _____
6.	data model	6. _____
7.	data storage system	7. _____
8.	non-relational database	8. _____
9.	semi-structured data	9. _____
10.	documentation	10. _____

〖Ex. 4〗 把下列单词翻译成英文。

1.	v. 访问，存取	1. _____
2.	n. 算法	2. _____
3.	v. 分析	3. _____
4.	v. 收集；获得	4. _____
5.	n. 数据集	5. _____
6.	v. 处理，应付	6. _____
7.	v. 使最优化	7. _____
8.	adj. 预测的，预言性的	8. _____
9.	v. 预加工，预处理	9. _____
10.	n. (计算机)软件	10. _____

〖Ex. 5〗 翻译句子。

1. It's the largest publicly available facial-recognition dataset.

2. You'll need big data technology to solve this problem.

3. This tool can be used in a variety of ways.

4. A few megabytes are required for the binaries, documentation and configuration files.

5. They have analyzed various factors.

6. The information is processed in a random order.

7. The server is designed to store huge amounts of data.

8. It is not possible to open the data in spreadsheet applications.

9. Most organizations cannot use such unstructured data efficiently.

10. All this information had to be input onto the computer.

Bigdata **Reading Material**

Top 9 Big Data Applications Across Industries

1. Banking and Securities

A study of 16 projects[①] in 10 top investment and retail banks shows that the challenges in this industry include: securities fraud early warning, tick[②] analytics, card fraud detection, archival of audit trails[③], enterprise credit risk reporting, trade visibility, customer data transformation, social analytics for trading, IT operations analytics, and IT policy compliance analytics, among others.

The Securities Exchange Commission (SEC[④]) is using big data to monitor financial market activity. They are currently using network analytics and natural language processors to catch illegal[⑤] trading activity in the financial markets.

Retail traders, Big banks, hedge funds[⑥], and other so-called "big boys" in the financial markets use big data for trade analytics used in high-frequency trading, pre-trade decision-support analytics[⑦], sentiment measurement, predictive analytics, etc.

This industry also heavily relies on big data for risk analytics, including; anti-money laundering[⑧], enterprise risk management, "Know Your Customer," and fraud mitigation[⑨].

---

① project ['prɒdʒekt]　n. 项目，工程

② tick [tɪk]　n. 赊账

③ audit trail：审计跟踪

④ SEC：证券交易委员会

⑤ illegal [ɪ'liːgl]　adj. 非法的

⑥ hedge fund：避险基金，对冲基金

⑦ pre-trade decision-support analytic：贸易前决策支持分析

⑧ anti-money laundering：反洗钱

⑨ mitigation [ˌmɪtɪ'geɪʃn]　n. 缓解，减轻

2. Communications, Media and Entertainment

Since consumers expect rich media[①] on-demand in different formats and a variety of devices, some big data challenges in the communications, media, and entertainment industry include:

- Collecting, analyzing, and utilizing consumer insights;
- Leveraging mobile and social media content;
- Understanding patterns of real-time, media content usage.

Organizations in this industry simultaneously analyze customer data along with behavioral data to create detailed customer profiles that can be used to:

- Create content for different target audiences;
- Recommend content on demand;
- Measure content performance.

A case in point is the Wimbledon Championships (YouTube Video) that leverages big data to deliver detailed sentiment analysis on the tennis matches to TV, mobile, and web users in real-time.

Spotify, an on-demand music service, uses Hadoop big data analytics, to collect data from its millions of users worldwide and then uses the analyzed data to give informed music recommendations to individual users.

Amazon Prime, which is driven to provide a great customer experience by offering video, music, and Kindle books in a one-stop-shop[②], also heavily utilizes big data.

3. Education

From a technical point of view, a significant challenge in the education industry is to incorporate big data from different sources and vendors and to utilize it on platforms that were not designed for the varying data.

From a practical point of view, staff and institutions have to learn new data management and analysis tools.

On the technical side, there are challenges to integrating data from different sources on different platforms and from different vendors that were not designed to work with one another.

Big data is used quite significantly in higher education. For example, University of Tasmania, a famous Australian university with over 30,000 students has deployed a Learning and Management System that tracks, among other things, when a student logs onto the system, how much time is spent on different pages in the system, as well as the overall progress of a student over time.

Big data is also used to measure teacher's effectiveness[③] to ensure a pleasant[①] experience

---

① rich media: 富媒体；多元媒体；多功能媒体

② one-stop-shop: 一站式

③ effectiveness [ɪˌfek'tɪvnɪs] n. 有效性；效益

for both students and teachers. Teacher's performance can be fine-tuned and measured against student numbers, subject matter, student demographics, student aspirations, behavioral classification, and several other variables.

4．Manufacturing and Natural Resources

The increasing demand for natural resources, including oil, agricultural products, minerals, gas, metals, and so on, has led to an increase in the volume, complexity, and velocity of data that is a challenge to handle.

Similarly, large volumes of data from the manufacturing industry are untapped[②]. The underutilization[③] of this information prevents the improved quality of products, energy efficiency, reliability, and better profit margins[④].

In the natural resources industry, big data allows for predictive modeling to support decision making that has been utilized for ingesting and integrating large amounts of data from geospatial[⑤] data, graphical data, text, and temporal data. Areas of interest where this has been used include seismic interpretation and reservoir characterization.

Big data has also been used in solving today's manufacturing challenges and to gain a competitive advantage, among other benefits.

5．Government

In governments, the most significant challenges are the integration and interoperability[⑥] of big data across different government departments and affiliated organizations.

In public services, big data has an extensive range of applications, including energy exploration, financial market analysis, fraud detection, health-related research, and environmental protection.

6．Insurance

Big data has been used in the industry to provide customer insights for transparent and simpler products, by analyzing and predicting customer behavior through data derived from social media, GPS-enabled devices, and CCTV footage. Big data also allows for better customer retention from insurance companies.

When it comes to claims management, predictive analytics from big data has been used to

---

① pleasant ['pleznt]　adj. 令人愉快的，友好的

② untapped [ˌʌn'tæpt]　adj. (有用或宝贵的东西)未开发的，未利用的

③ underutilization [ˌʌndəˌjuːtəlaɪz'eɪʃn]　n. 利用不足，未充分使用

④ profit margin：利润率

⑤ geospatial [ˌdʒiːəʊ'speɪʃəl]　adj. 地理空间的

⑥ interoperability ['ɪntərɒpərə'bɪləti]　n. 互用性，协同工作的能力

offer faster service since massive amounts of data can be analyzed mainly in the underwriting stage. Fraud detection has also been enhanced.

Through massive data from digital channels and social media, real-time monitoring of claims throughout the claims cycle has been used to provide insights.

7. Retail and Wholesale Trade

From traditional brick and mortar retailers[①] and wholesalers to current day e-commerce traders, the industry has gathered a lot of data over time. This data, derived from customer loyalty cards, POS scanners, RFID[②], etc. are used to improve customer experiences on the whole.

8. Transportation

Some applications of big data by governments, private organizations, and individuals in transportation include:

• Governments use of big data: traffic control, route planning, intelligent transport systems, congestion[③] management (by predicting traffic conditions).

• Private-sector use of big data in transport: revenue management, technological enhancements, logistics and for competitive advantage (by consolidating[④] shipments and optimizing freight movement).

• Individual use of big data includes route planning[⑤] to save on fuel and time, for travel arrangements in tourism, etc.

9. Energy and Utilities

Smart meter[⑥] readers allow data to be collected almost every 15 minutes as opposed to once a day with the old meter readers. This granular data is being used to analyze the consumption of utilities better, which allows for improved customer feedback and better control of utilities use.

In utility companies, the use of big data also allows for better asset and workforce management, which is useful for recognizing errors and correcting them as soon as possible before complete failure is experienced.

---

① brick and mortar retailer：实体店

② Radio Frequency Identification 的缩写，射频识别技术

③ congestion [kən'dʒestʃən] n. 拥挤，堵车

④ consolidate [kən'sɒlɪdeɪt] v. 加强，巩固；合并

⑤ route planning：线路规划

⑥ smart meter：智慧电表

Unit 2

Database Basics

1. Database

It is often abbreviated DB. It is a collection of information organized in such a way that a computer program can quickly select desired pieces of data. You can think of a database as an electronic filing system.

Traditional databases are organized by fields, records and files. A field is a single piece of information; a record is one complete set of fields; and a file is a collection of records. For example, a telephone book is analogous to a file. It contains a list of records, each of which consists of three fields: name, address and telephone number.

An alternative concept in database design is known as Hypertext. In a Hypertext database, any object, whether it be a piece of text, a picture, or a film, can be linked to any other object. Hypertext databases are particularly useful for organizing large amounts of disparate information, but they are not designed for numerical analysis.

To access information from a database, you need a database management system (DBMS). This is a collection of programs that enables you to enter, organize, and select data in a database.

Increasingly, the term database is used as shorthand for database management system.

2. Field

It is a space allocated for a particular item of information. A tax form, for example, contains a number of fields: one for your name, one for your Social Security number, one for your income, and so on. In database systems, fields are the smallest units of information you can access. In spreadsheets, fields are called cells.

Most fields have certain attributes associated with them. For example, some fields are numeric whereas others are textual, some are long, while others are short. In addition, every field has a name, called the field name.

In database management systems, a field can be required, optional, or calculated. A required field is one in which you must enter data, while an optional field is one you may leave blank. A calculated field is one whose value is derived from some formula involving other fields. You do not enter data into a calculated field; the system automatically determines the correct value.

A collection of fields is called a record.

3. Record

In database management systems, it is a complete set of information. Records are composed of fields, each of which contains one item of information. A set of records constitutes a file. For example, a personnel file might contain records that have three fields: a name field, an address field and a phone number field.

In relational database management systems, records are called tuples.

Some programming languages allow you to define a special data structure called a record. Generally, a record is a combination of other data objects. For example, a record might contain three integers, a floating-point number and a character string.

4. Database Management System

It is a collection of programs that enables you to store, modify, and extract information from a database. There are many different types of DBMSs, ranging from small systems that run on personal computers to huge systems that run on mainframes. The following are examples of database applications:

- computerized library systems;
- automated teller machines;
- flight reservation systems;
- computerized parts inventory systems.

From a technical standpoint, DBMSs can differ widely. The terms relational, network, flat, and hierarchical all refer to the way a DBMS organizes information internally. The internal organization can affect how quickly and flexibly you can extract information.

Requests for information from a database are made in the form of a query, which is a stylized question. For example, the query

SELECT ALL WHERE NAME = "SMITH" AND AGE > 35

requests all records in which the NAME field is SMITH and the AGE field is greater than 35. The set of rules for constructing queries is known as a query language. Different DBMSs support different query languages, although there is a semi-standardized query language called SQL (Structured Query Language). Sophisticated languages for managing database systems are called fourth-generation languages, or 4GLs for short.

The information from a database can be presented in a variety of formats. Most DBMSs include a report writer program that enables you to output data in the form of a report. Many DBMSs also include a graphics component that enables you to output information in the form of graphs and charts.

5. SQL

It is the abbreviation of structured query language, and is pronounced either see-kwell or as separate letters. SQL is a standardized query language for requesting information from a database. The original version called SEQUEL (structured English query language) was designed by an IBM research center in 1974 and 1975. SQL was first introduced as a commercial database system in 1979 by Oracle Corporation.

Historically, SQL has been the favorite query language for database management systems running on minicomputers and mainframes. Increasingly, however, SQL is being supported by PC database systems because it supports distributed databases (databases that are spread out over several computer systems). This enables several users on a local-area network to access the same database simultaneously.

Although there are different dialects of SQL, it is nevertheless the closest thing to a standard query language that currently exists. In 1986, ANSI approved a rudimentary version of SQL as the official standard, but most versions of SQL since then have included many extensions to the ANSI standard. In 1991, ANSI updated the standard. The new standard is known as SAG SQL.

6. Distributed Database

It is a database that consists of two or more data files located at different sites on a computer network. Because the database is distributed, different users can access it without interfering with one another. However, the DBMS must periodically synchronize the scattered databases to make sure that they all have consistent data.

New Words

database	['deɪtəbeɪs]	n. 数据库
field	[fiːld]	n. 字段
hypertext	['haɪpətekst]	n. 超文本
particularly	[pə'tɪkjələli]	adv. 独特地，显著地
disparate	['dɪspərət]	adj. 异类的，不同的
increasingly	[ɪn'kriːsɪŋli]	adv. 越来越(多)地，逐渐增加地，日益地
shorthand	['ʃɔːthænd]	n. 速记
allocate	['æləkeɪt]	v. 分派，分配
textual	['tekstʃuəl]	adj. 本文的，原文的
required	[rɪ'kwaɪəd]	adj. 必需的
formula	['fɔːmjələ]	n. 公式，规则
geographical	[ˌdʒiːə'græfɪkl]	adj. 地理学的，地理的

representative	[ˌreprɪ'zentətɪv]	n. 代表
		adj. 典型的，有代表性的
tuple	[tʌpl]	n. 元组
integer	['ɪntɪdʒə]	n. 整型数，整型数
string	[strɪŋ]	n. 串
extract	['ekstrækt]	v. 提取，析取
huge	[hjuːdʒ]	adj. 巨大的，极大的，无限的
mainframe	['meɪnfreɪm]	n. 主机，大型机
flight	[flaɪt]	n. 飞机的航程，班机
reservation	[ˌrezə'veɪʃn]	n. 保留，预定，预约
standpoint	['stændpɔɪnt]	n. 立场，观点，角度
hierarchical	[ˌhaɪə'rɑːkɪkl]	adj. 分等级的
request	[rɪ'kwest]	v.&n. 请求
query	['kwɪəri]	n.&v. 查询
sophisticated	[sə'fɪstɪkeɪtɪd]	adj. 成熟的，富有经验的；复杂的；精致的
minicomputer	['mɪnikəmpjuːtə]	n. 小型机，小型计算机
simultaneously	[ˌsɪməl'teɪnɪəsli]	adv. 同时地
rudimentary	[ˌruːdɪ'mentri]	adj. 基本的，初步的
periodically	[ˌpɪərɪ'ɒdɪkli]	adv. 周期性地，定时性地
synchronize	['sɪŋkrənaɪz]	v. 同步
scattered	['skætəd]	adj. 离散的，分散的
consistent	[kən'sɪstənt]	adj. 一致的，调和的，相容的

Phrases

database management system (DBMS)	数据库管理系统
field name	字段名
data structure	数据结构
floating-point number	浮点数
character string	字符串
distributed database	分布式数据库
spread out	展开
official standard	官方标准

Abbreviations

SQL (Structured Query Language)	结构化查询语言

| 4GL (Fourth-Generation Languages) | 第四代语言 |
| SEQUEL (Structured English QUEry Language) | 结构化英语查询语言 |

 ## Analysis of Difficult Sentences

[1] It is a collection of information organized in such a way that a computer program can quickly select desired pieces of data.

本句中，organized in such a way that a computer program can quickly select desired pieces of data 是一个过去分词短语，作定语，修饰和限定 a collection of information。该短语可以扩展为一个定语从句 which is organized in such a way that a computer program can quickly select desired pieces of data。在该从句中 in such a way 作方式状语，修饰 organized，that a computer program can quickly select desired pieces of data 是一个定语从句，修饰和限定 a way

[2] A required field is one in which you must enter data, while an optional field is one you may leave blank.

本句中，in which you must enter data 是一个介词前置的定语从句，修饰和限定它前面的 one，one 指 A required field。you may leave blank 是一个定语从句，修饰和限定它前面的 one，这个 one 指 an optional field。while 表示对比。

[3] Records are composed of fields, each of which contains one item of information.

本句中，each of which contains one item of information 是一个非限定性定语从句，对 fields 进行补充说明。

[4] Although there are different dialects of SQL, it is nevertheless the closest thing to a standard query language that currently exists.

本句中，Although there are different dialects of SQL 是一个让步状语从句，修饰谓语 is nevertheless the closest thing。that currently exists 是一个定语从句，修饰和限定 a standard query language。

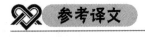 参考译文

数 据 库 基 础

1. 数据库

数据库通常简写为 DB。数据库是一个信息集合，以一种计算机程序能快速选择所需数据的方式组织而成。可以将数据库想象为电子文档系统。

传统数据库由字段、记录和文件组成。字段是一条信息；记录是一组完整的字段；文档是记录的集合。例如，电话本和文件类似。它包括含记录的列表，每个记录包含三个字

段：姓名、地址和电话号码。

数据库设计的另一个概念是超文本。在超文本数据库中，任何对象，无论是文本、图像还是电影，都可以链接到任意其他对象。在组织大量不同的信息时超文本数据库尤其有用，但是它们不是为数字分析设计的。

要通过数据库访问信息，就需要数据库管理系统(DBMS)。这是一套允许进入、组织和选择数据库中数据的程序。

数据库这个术语越来越多地被用成数据库管理系统的简略表达方式。

2. 字段

它是分配特定信息项的空间。例如，纳税申报单包括许多字段：一个是姓名、一个是社会保险号码、一个是收入，等等。在数据库系统中，字段是能访问的最小的信息单元。在电子表格中，字段被称为单元。

大多数字段有其特有属性。比如，有些字段是数字的，而有些是文本的，有些长，有些短。此外，每个字段都有名字，被称为字段名。

在数据库管理系统中，字段可以是必填字段、可选字段或计算字段。必填字段是必须在其中输入数据的字段；可选字段是可以留空的字段；计算字段是指其值由涉及其他字段的公式推导而来的字段。不用在计算字段中输入数据；系统会自动确定正确的值。

字段的集合叫作记录。

3. 记录

在数据库管理系统中，记录是一个完整的信息集合。记录由字段组成，每个字段包含一个信息项。一组记录组成一个文件。例如，职员文件可能包括有三个字段的记录：姓名字段、地址字段和电话号码字段。

在关系数据库管理系统中，记录被称为元组。

有些程序语言允许定义特殊的数据结构，称为记录。通常一个记录是其他数据对象的组合。例如，记录可能包含三个整数、一个浮点数和一个字符串。

4. 数据库管理系统

数据库管理系统是一个程序集合，允许存储、修改和提取数据库中的信息。数据库管理系统有很多种，从运行在个人电脑上的小型系统到运行在大型机上的巨型系统。下面是数据库应用的几个例子：

- 计算机化的图书馆系统；
- 自动取款机；
- 航班预订系统；
- 计算机化零件库存系统。

从技术的角度来看，不同的数据库管理系统相差很大。术语关系、网络、平台和等级都是指数据库组织内部信息的方式。内部组织会影响到提取信息的速度和灵活性。

以查询的形式对数据库提出信息请求是程式化的问题。例如，查询

SELECT ALL WHERE NAME = "SMITH" AND AGE > 35

请求所有 NAME 字段为 SMITH 和年龄字段大于 35 的记录。构成查询的规则集叫查询语言。尽管有一个名为 SQL(结构化查询语言)的半标准查询语言，但不同的数据库管理系统支持不同的查询语言。用于管理数据库系统的成熟语言称为第四代语言，简称 4GLs。

数据库中的信息可以以多种格式显示。大部分数据库管理系统有一个报告编写程序，可以用报告的形式输出数据。许多数据库管理系统还有一个图形部件，可以用图形和图表的形式输出信息。

5. SQL

SQL 是结构化查询语言的缩写，发音为 see-kwell 或按单个字母发音。SQL 是在数据库中请求信息的标准查询语言。原版本名为 SEQUEL(结构化英语查询语言)，由 IBM 研究中心于 1974 和 1975 年研制。1979 年，SQL 首先由甲骨文公司作为商业数据库系统引入。

历史上，SQL 曾是在小型机和大型机上运行的十分受欢迎的数据库管理系统查询语言。然而，如今 SQL 越来越被 PC 数据库系统支持，因为它支持分布式数据库(分布在多个计算机系统上的数据库)。这就允许局域网中多个用户同时访问同一个数据库。

尽管 SQL 有很多版本，它仍然是现有最接近标准查询语言的语言。1986 年 ANSI 批准一个 SQL 的基本版本作为官方标准，但是从那时起，大多数 SQL 版本包含了 ANSI 标准的众多扩展。1991 年，ANSI 升级了该标准。新的标准名为 SAG SQL。

6. 分布式数据库

分布式数据库是一种在计算机网络的不同站点包含两个或多个数据文件的数据库。由于数据库是分布式的，不同用户可以在互不影响的情况下进行访问。但是，数据库管理系统必须定期同步分散的数据库，确保数据一致。

Bigdata Text B

What Is a Data Warehouse?

1. Definition of a Data Warehouse

A data warehouse is a type of data management system that is designed to enable and support business intelligence (BI) activities, especially analytics. Data warehouses are solely intended to perform queries and analysis and often contain large amounts of historical data. The data within a data warehouse is usually derived from a wide range of sources such as application log files and transaction applications.

A data warehouse centralizes and consolidates large amounts of data from multiple sources. Its analytical capabilities allow organizations to derive valuable business insights from their data to improve decision-making. Over time, it builds a historical record that can be invaluable to data scientists and business analysts. Because of these capabilities, a data warehouse can be considered an organization's "single source of truth".

A typical data warehouse often includes the following elements:

- A relational database to store and manage data;
- An extraction, loading and transformation solution for preparing the data for analysis;
- Statistical analysis, reporting and data mining capabilities;
- Client analysis tools for visualizing and presenting data to business users;
- Other more sophisticated analytical applications that generate actionable information by applying data science and artificial intelligence (AI) algorithms, or graph and spatial features that enable more kinds of analysis of data at scale.

2. Benefits of a Data Warehouse

Data warehouses offer the overarching and unique benefit of allowing organizations to analyze large amounts of variant data and extract significant value from it as well as to keep a historical record.

Four unique characteristics (described by computer scientist William Inmon, who is considered the father of the data warehouse) allow data warehouses to deliver this overarching benefit. According to this definition, data warehouses are:

- Subject-oriented. They can analyze data about a particular subject or functional area (such as sales).
- Integrated. Data warehouses create consistency among different data types from disparate sources.
- Nonvolatile. Once data is in a data warehouse, it's stable and doesn't change.
- Time-variant. Data warehouse analysis looks at change over time.

A well-designed data warehouse will perform queries very quickly, deliver high data throughput, and provide enough flexibility for end users to "slice and dice" or reduce the volume of data for closer examination to meet a variety of demands—whether at a high level or at a very fine, detailed level. The data warehouse serves as the functional foundation for middleware BI environments that provide end users with reports, dashboards, and other interfaces.

3. Data Warehouse Architecture

A data warehouse architecture defines the arrangement of the data in different databases. A traditional data warehouse requires on-premise servers for all components of the warehouse to function.

When designing a corporation's data warehouse, there are three types of traditional data warehouse models to consider:

1) Single-tier data warehouse architecture

A single-tier data warehouse architecture centers on producing a dense set of data and reducing the volume of data deposited. Although it is beneficial for eliminating redundancies, this type of warehouse architecture is not suitable for businesses with complex data requirements and numerous data streams. This is where multi-tier data warehouse architectures come in as they deal with more complex data streams.

2) Two-tier data warehouse architecture

In comparison, a two-tier data warehouse architecture splits the tangible data sources from the warehouse itself. Unlike a single-tier, the two-tier architecture uses a system and a database server. This is most commonly used in small organizations where a server is used as a data mart. Although it is more efficient at data storage and organization, the two-tier architecture is not scalable.

3) Three-tier data warehouse architecture

The three-tier data warehouse architecture is the most widely use type of modern data warehouse architecture as it produces a well-organized data flow from raw information to valuable insights. It consists of three tiers, namely the top, middle and bottom.

The bottom tier in the data warehouse architecture typically comprises the databank server that creates an abstraction layer on data from numerous sources, like transactional databanks utilized for front-end uses.

The middle tier includes an Online Analytical Processing (OLAP) server. From a user's perspective, this level alters the data into an arrangement that is more suitable for analysis and multifaceted probing. Since it includes an OLAP server pre-built in the architecture, we can also call it the OLAP-focused data warehouse.

The top tier is the client level. It includes the tools and Application Programming Interface (API) that you connect and get data out from the data warehouse. It can be tools that are used for high-level data analysis, inquiring, and reporting.

4. Cloud Data Warehouse

A cloud data warehouse uses the cloud to ingest and store data from disparate data sources. It offers the same characteristics and benefits of on-premises data warehouses but with the added benefits of cloud computing. Cloud data warehouses allow enterprises to focus solely on extracting value from their data rather than having to build and manage the hardware and software infrastructure to support the data warehouse.

Some of the advantages of cloud data warehouses include:

- Elastic, scale-out support for large or variable compute or storage requirements;
- Ease of use;
- Ease of management;
- Cost savings.

The best cloud data warehouses are fully managed and self-driving, ensuring that even beginners can create and use a data warehouse with only a few clicks. An easy way to start your migration to a cloud data warehouse is to run your cloud data warehouse on-premises, behind your data center firewall which complies with data sovereignty and security requirements.

In addition, most cloud data warehouses follow a pay-as-you-go model, which brings added cost savings to customers.

5. Modern Data Warehouse

Whether they're part of IT, data engineering, business analytics, or data science teams, different users across the organization have different needs for a data warehouse.

A modern data architecture addresses those different needs by providing a way to manage all data types, workloads and analysis. It consists of architecture patterns with necessary components integrated to work together in alignment with industry best practices. The modern data warehouse includes:

• A converged database that simplifies management of all data types and provides different ways to use data;

- Self-service data ingestion and transformation services;
- Support for SQL, machine learning, graph, and spatial processing;
- Multiple analytics options that make it easy to use data without moving it;
- Automated management for simple provisioning, scaling, and administration.

A modern data warehouse can efficiently streamline data workflows in a way that other warehouses can't. This means that everyone, from analysts and data engineers to data scientists and IT teams, can perform their jobs more effectively and pursue the innovative work that moves the organization forward, without countless delays and complexity.

6. Designing a Data Warehouse

When an organization sets out to design a data warehouse, it must begin by defining its specific business requirements, agreeing on the scope, and drafting a conceptual design. The organization can then create both the logical and physical design for the data warehouse. The logical design involves the relationships between the objects, and the physical design involves the best way to store and retrieve the objects. The physical design also incorporates transportation, backup, and recovery processes.

Any data warehouse design must address the following:

- Specific data content;
- Relationships within and between groups of data;
- The systems environment that will support the data warehouse;
- The types of data transformations required;
- Data refresh frequency.

A primary factor in the design is the needs of the end users. Most end users are interested in performing analysis and looking at data in aggregate, instead of as individual transactions. However, often end users don't really know what they want until a specific need arises. Thus, the planning process should include enough exploration to anticipate needs. Finally, the data warehouse design should allow room for expansion and evolution to keep pace with the evolving needs of end users.

New Words

contain	[kən'teɪn]	v. 包含，容纳
centralize	['sentrəlaɪz]	v. 使中心化，集中化
multiple	['mʌltɪpl]	adj. 多重的；多个的
invaluable	[ɪn'væljʊəbl]	adj. 非常宝贵的，无法估计的
element	['elɪmənt]	n. 要素，元素
client	['klaɪənt]	n. 客户
actionable	['ækʃənəbl]	adj. 可行动的
spatial	['speɪʃl]	adj. 空间的
variant	['veərɪənt]	adj. 变化的，不同的，相异的
		n. 变量
subject-oriented	['sʌbdʒɪkt 'ɔːrɪentɪd]	adj. 面向主题的
consistency	[kən'sɪstənsi]	n. 连贯性，一致性
nonvolatile	['nɒn'vɒlətaɪl]	adj. 非易失性的
stable	['steɪbl]	adj. 稳定的；牢固的
throughput	['θruːpʊt]	n. 吞吐量；流率
foundation	[faʊn'deɪʃn]	n. 基础
middleware	['mɪdlweə]	n. 中间件；中间设备
dashboard	['dæʃbɔːd]	n. 仪表板，仪表盘
interface	['ɪntəfeɪs]	n. 界面；接口
arrangement	[ə'reɪndʒmənt]	n. 安排；布置
dense	[dens]	adj. 密集的；浓的
deposit	[dɪ'pɒzɪt]	v. 放下；存放；使沉淀
tangible	['tændʒəbl]	adj. 有形的；可触知的
databank	['deɪtəbæŋk]	n. 数据库
perspective	[pə'spektɪv]	n. 观点，看法
alter	['ɔːltə]	v. 改变，更改；转换
multifaceted	[,mʌltɪ'fæsɪtɪd]	adj. 多方面的
inquiry	[ɪn'kwaɪəri]	n. 查询；调查，审查；询问
ingest	[ɪn'dʒest]	v. 采集，获取，吸收
elastic	[ɪ'læstɪk]	adj. 有弹性的；灵活的
click	[klɪk]	v. (用鼠标)点击
firewall	['faɪəwɔːl]	n. 防火墙
workload	['wɜːkləʊd]	n. 工作量，工作负担
converge	[kən'vɜːdʒ]	v. 汇集，聚集
ingestion	[ɪn'dʒestʃən]	n. 摄取；采集

automate	[ˈɔːtəmeɪt]	v. 使自动化
streamline	[ˈstriːmlaɪn]	v. 组织；使现代化；使简单化
pursue	[pəˈsjuː]	v. 追求；进行
delay	[dɪˈleɪ]	n. 延迟，延期，耽误
complexity	[kəmˈpleksəti]	n. 复杂性
draft	[drɑːft]	v. 起草
conceptual	[kənˈseptʃʊəl]	adj. 概念的，观念的
object	[ˈɒb.dʒɪkt]	n. 对象；物体；目标
relationship	[rɪˈleɪʃnʃɪp]	n. 关系
frequency	[ˈfriːkwənsi]	n. 频率，次数
aggregate	[ˈægrɪgɪt]	n. 总数，合计
		adj. 总数的，总计的
		v. 总计，汇集
anticipate	[ænˈtɪsɪpeɪt]	v. 预期；预计

Phrases

data warehouse	数据仓库
data management system	数据管理系统
historical data	历史数据
derive from	采自，来自
log file	日志文件
on-premise server	内部服务器
single-tier data warehouse	单层数据仓库
data stream	数据流
multi-tier data warehouse	多层数据仓库
two-tier data warehouse	两层数据仓库
three-tier data warehouse	三层数据仓库
raw information	原生信息，原始资料
data sovereignty	数据主权
security requirement	安全要求
cloud data warehouse	云数据仓库
pay-as-you-go model	按需支付的模型
in alignment with …	与……一致
set out	出发；启程；开始工作
logical design	逻辑设计
physical design	物理设计
data refresh	数据刷新，数据更新

end user	最终用户，终端用户
keep pace with…	跟上……步伐

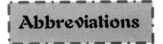

API (Application Programming Interface)   应用程序接口

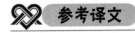

什么是数据仓库？

1. 数据仓库的定义

数据仓库是一种数据管理系统，旨在启动和支持商业智能(BI)活动，尤其是智能分析。数据仓库仅用于执行查询和分析，通常包含大量历史数据。数据仓库中的数据通常来源广泛，例如来自应用程序日志文件和事务应用程序。

数据仓库集中并整合多个来源的大量数据。它的分析能力使组织能够从数据中获得有价值的业务见解，以改进决策。随着时间的推移，它会建立对数据科学家和业务分析师来说非常宝贵的历史记录。基于这些功能，数据仓库可以被视为组织的"单一真实来源"。

典型的数据仓库通常包括以下元素：
- 用于存储和管理数据的关系数据库；
- 用于准备分析数据的提取、加载和转换 (ELT) 解决方案；
- 统计分析、报告和数据挖掘功能；
- 用于向业务用户可视化呈现数据的客户端分析工具；
- 其他更复杂的分析应用程序，通过应用数据科学和人工智能(AI)算法生成可操作的信息或图形和实现更大规模的数据分析的空间特征。

2. 数据仓库的好处

数据仓库提供了整体和独特的好处，即允许组织分析大量可变数据并从中提取重要价值并保存历史记录。

四个独有的特征(由被誉为是数据仓库之父的计算机科学家 William Inmon 描述)使数据仓库能够提供这一整体优势。根据该定义，数据仓库是：
- 面向主题：可以分析特定主题或功能区域(例如销售)的数据。
- 整合的：数据仓库在来自不同来源的不同数据类型之间创建一致性。
- 非易失性：一旦数据进入数据仓库，它就是稳定的并且不会改变。
- 时变：数据仓库分析着眼于随时间的变化。

一个设计良好的数据仓库将非常快速地执行查询、提供高数据吞吐量，并为最终用户提供足够的灵活性来"切片和切块"或减少数据量以进行仔细检查来满足各种需求——无论是在高级别还是非常精细、详细的级别。数据仓库作为中间件商业智能环境的功能基础，为最终用户提供报告、仪表板和其他界面。

3. 数据仓库架构

数据仓库架构定义了在不同数据库中数据的排列。传统的数据仓库需要内部部署的服务器才能使仓库的所有组件正常运行。

在设计公司的数据仓库时，需要考虑三种类型的传统数据仓库模型：

1) 单层数据仓库架构

单层数据仓库架构的核心是生成密集的数据集和减少存储的数据量。这种仓库架构虽然有利于消除冗余，但不适合数据需求复杂、数据流众多的业务。这就是多层数据仓库架构在处理更复杂的数据流时发挥作用的地方。

2) 两层数据仓库架构

相比之下，两层数据仓库架构将有形数据源与仓库本身分开。与单层不同，两层架构使用系统和数据库服务器。这最常用于将服务器用作数据集市的小型组织。虽然两层架构在数据存储和组织方面更高效，但不可扩展。

3) 三层数据仓库架构

三层数据仓库架构是现代数据仓库架构中使用最广泛的类型，因为它产生了从原始信息到有价值的见解的组织良好的数据流。它由三层组成，即顶层、中间层和底层。

数据仓库架构的底层通常包括数据库服务器，该服务器为众多来源的数据创建抽象层，例如用于前端使用的事务数据库。

中间层包括在线分析处理(OLAP)服务器。从用户的角度来看，这一层将数据更改为更适合分析和多方面探索的排列。由于它包含了在架构中预先构建的 OLAP 服务器，我们也可以将其称为以 OLAP 为中心的数据仓库。

顶层是客户端级别。它包括连接并从数据仓库中获取数据的工具和应用程序编程接口(API)，可用于高级数据分析、查询和报告。

4. 云数据仓库

云数据仓库使用云从不同的数据源采集和存储数据。它可提供与本地数据仓库相同的特性和优势，但具有云计算的额外优势。云数据仓库允许企业专注于从其数据中提取价值，而不必构建和管理硬件及软件基础设施来支持数据仓库。

云数据仓库的一些优势包括：
- 对大型或可变计算或存储要求提供弹性和横向扩展的支持；
- 便于使用；
- 易于管理；
- 节约成本。

最好的云数据仓库是完全托管和自动驱动的，即使是初学者也只需点击几下鼠标即可创建和使用数据仓库。开始迁移到云数据仓库的一种简单方法是在符合数据主权和安全要求的数据中心防火墙后面在本地运行云数据仓库。

此外，大多数云数据仓库遵循即用即付模式，这为客户节约额外的成本。

5．现代数据仓库

无论是 IT、数据工程、业务分析或数据科学团队的一部分，组织中的不同用户对数据仓库有不同的需求。

现代数据架构通过提供一种管理所有数据类型、工作负载和分析的方法来满足这些不同的需求。现代数据仓库由集成了必要组件的架构模式组成，以便与行业最佳实践保持一致。现代数据仓库包括：

- 融合数据库，可简化所有数据类型的管理并提供不同的数据使用方式；
- 自助式数据采集和转换服务；
- 支持 SQL、机器学习、图形和空间处理；
- 多种分析选项，无需移动数据即可轻松使用数据；
- 用于简单配置、扩展和管理的自动化管理。

现代数据仓库能够以其他仓库无法做到的方式有效地简化数据工作流。这意味着从分析师和数据工程师到数据科学家和 IT 团队的每个人，都可以更有效地执行工作，并追求推动组织向前发展的创新工作，而不会出现无数的延迟性和复杂性。

6．设计数据仓库

当一个组织开始设计数据仓库时，必须首先定义其特定的业务需求，就范围达成一致，并起草概念设计。然后，组织可以为数据仓库创建逻辑和物理设计。逻辑设计涉及对象之间的关系，物理设计涉及存储和检索对象的最佳方式。物理设计还包含传输、备份和恢复过程。

任何数据仓库设计都必须解决以下问题：

- 具体数据内容；
- 数据组内部及其之间的关系；
- 支持数据仓库的系统环境；
- 所需的数据转换类型；
- 数据刷新频率。

设计中的一个主要因素是最终用户的需求。大多数最终用户对执行分析和查看汇总数据感兴趣，而不是单个的业务。然而，在出现特定需求之前，最终用户通常并不真正知道他们想要什么。因此，规划过程应该包括足够的探索以预测需求。最后，数据仓库设计应该留有扩展和演变的空间，以跟上最终用户不断变化的需求。

Bigdata **Exercises**

〖Ex. 1〗 根据 Text A 回答以下问题。

1. What is database? How are traditional database organized?

2. What is a field? What is a record? And what is a file?

3. What are Hypertext databases particularly useful for?

4. What do you need to access information from a database? What is it?

5. What are fields in database systems? What are they called in spreadsheets?

6. What is the difference between a required field, an optional field and a calculated field?

7. What is record in database management systems? What is the relationship between records and a file?

8. What are examples of database applications?

9. What is SQL?

10. What is a distributed database?

〖Ex. 2〗 根据 Text B 回答以下问题。

1. What is a data warehouse?

2. What does a data warehouse do?

3. What do data warehouses offer?

4. What are the four unique characteristics of data warehouses?

5. How many types of traditional data warehouse models to consider when designing a corporation's data warehouse? What are they?

6. What do some of the advantages of cloud data warehouses include?

7. How does a modern data architecture addresses those different needs?

8. What can a modern data warehouse efficiently do?

9. When an organization sets out to design a data warehouse, how must it begin?

10. What must any data warehouse design address?

〖Ex. 3〗 把下列词组翻译成中文。

1. data structure   1. _____

2. distributed database   2. _____

3. field name   3. _____

4. cloud data warehouse   4. _____

5. character string   5. _____

6. data management system   6. _____

7. data warehouse   7. _____

8. data stream   8. _____

9. log file   9. _____

10. raw information   10. _____

〖Ex. 4〗 把下列单词翻译成英文。

1. n. 数据库　　　　　　　　　1. _____
2. v. 提取，析取　　　　　　　2. _____
3. n. 字段，字段　　　　　　　3. _____
4. n. 主机，大型机　　　　　　4. _____
5. v. 同步　　　　　　　　　　5. _____
6. n. 复杂性　　　　　　　　　6. _____
7. n. 查询；调查，审查；询问　7. _____
8. n. 界面；接口　　　　　　　8. _____
9. n. 中间件；中间设备　　　　9. _____
10. adj. 非易失性的　　　　　　10. _____

〖Ex. 5〗 翻译句子。

1. Computer databases typically contain aggregations of data records or files.
2. The relational database was invented by E. F. Codd at IBM in 1970.
3. SQL is the standard user and application program interface to a relational database.
4. SQL statements are used both for interactive queries for information from a relational database and for gathering data for reports.
5. Data independence makes stored data as accessible as possible.
6. Deleting tells the computer that data or file is no longer need.
7. A deleted file usually remains stored until the operating system reuses the space containing the deleted material.
8. A database management system allows users to control security and data integrity requirements.
9. A database management system handles user requests for database action.
10. A database is a collection of information that is organized so that it can easily be accessed, managed and updated.

Bigdata Reading Material

Data Mart

In a market dominated by big data and analytics, data marts① are one key to efficiently

---

① data mart：数据集市

transforming information into insights. Data warehouses typically deal with large data sets, but data analysis requires easy-to-find and readily[①] available data. Should a business person have to perform complex queries just to access the data they need for their reports? No—and that's why smart companies use data marts.

A data mart is a subject-oriented database that is often a partitioned segment[②] of an enterprise data warehouse. The subset[③] of data held in a data mart typically aligns with a particular business unit like sales, finance, or marketing. Data marts accelerate[④] business processes by allowing access to relevant information in a data warehouse or operational data store within days, as opposed to months or longer. Because a data mart only contains the data applicable to a certain business area, it is a cost-effective way to gain actionable insights quickly.

1. Data Mart vs. Data Warehouse

Data marts and data warehouses are both highly structured repositories where data is stored and managed until it is needed. However, they differ in the scope of data stored: data warehouses are built to serve as the central store of data for the entire business, whereas a data mart fulfills[⑤] the request of a specific division or business function. Because a data warehouse contains data for the entire company, it is best practice to have strictly control on who can access it. Additionally, querying the data you need in a data warehouse is an incredibly difficult task for the business. Thus, the primary purpose of a data mart is to isolate[⑥]—or partition—a smaller set of data from a whole to provide easier data access for the end consumers (see table 2-1).

Table 2-1　Date Mart vs. Date Ware house

Type	Data Mart	Data Warehouse
Size	< 100 GB	100 GB +
Subject	Single Subject	Multiple Subjects
Scope	Line-of-Business	Enterprise-wide
Data Sources	Few Sources	Many Source Systems
Data Integration	One Subject Area	All Business Data
Time to Build	Minutes, Weeks, Months	Many Months to Years

A data mart can be created from an existing data warehouse—the top-down approach[⑦]—or

① readily ['redɪli]　adv. 快捷地；轻而易举地；便利地

② segment ['seɡmənt]　n. 部分，段落

③ subset ['sʌbset]　n. 子集

④ accelerate [ək'seləreɪt]　v. 加速，加快

⑤ fulfill [fʊl'fɪl]　vt. 执行

⑥ isolate ['aɪsəleɪt]　vi. 隔离，孤立

⑦ top-down approach：自顶向下法

from other sources, such as internal operational systems or external data. Similar to a data warehouse, it is a relational database that stores transactional data (time value, numerical order, reference to one or more object) in columns and rows making it easy to organize and access.

On the other hand, separate business units may create their own data marts based on their own data requirements. If business needs dictate, multiple data marts can be merged together to create a single, data warehouse. This is the bottom-up development approach.

2. Types of Data Marts

There are three types of data marts: dependent, independent[①], and hybrid[②]. They are categorized based on their relation to the data warehouse and the data sources that are used to create the system.

2.1 Dependent data marts

A dependent data mart is created from an existing enterprise data warehouse. It is the top-down approach that begins with storing all business data in one central location, then extracts a clearly defined portion of the data when needed for analysis.

To form a data mart, a specific set of data is aggregated (formed into a cluster) from the warehouse, restructured[③], then loaded to the data mart where it can be queried. It can be a logical view or physical subset of the data warehouse:

• Logical view—A virtual table/view that is logically—but not physically—separated from the data warehouse.

• Physical subset—Data extract that is a physically separate database from the data warehouse.

2.2 Independent data marts

An independent data mart is a stand-alone[④] system—created without the use of a data warehouse—that focuses on one subject area or business function. Data is extracted from internal or external data sources (or both), processed, then loaded to the data mart repository where it is stored until needed for business analytics.

Independent data marts are not difficult to design and develop. They are beneficial to achieve short-term goals but may become cumbersome[⑤] to manage—each with its own ETL tool and logic—as business needs expand and become more complex.

2.3 Hybrid Data Marts

A hybrid data mart combines data from an existing data warehouse and other operational

---

① independent [ˌɪndɪˈpendənt]　adj. 独立的，自立的，无关联的

② hybrid [ˈhaɪbrɪd]　adj. 混合的

③ restructure [ˌriːˈstrʌktʃə]　v. 重建，调整，重组

④ stand-alone [stænd əˈləʊn]　adj. 独立的

⑤ cumbersome [ˈkʌmbəsəm]　adj. 麻烦的

source systems. It unites the speed and end-user focus of a top-down approach with the benefits of the enterprise-level integration of the bottom-up method[①].

3. Structure of a Data Mart

Similar to a data warehouse, a data mart may be organized using a star, snowflake, vault, or other schema as a blueprint[②]. IT teams typically use a star schema consisting of one or more fact tables (set of metrics relating to a specific business process or event) referencing dimension tables (primary key joined to a fact table) in a relational database.

The benefit of a star schema is that fewer joins are needed when writing queries, as there is no dependency between dimensions. This simplifies the ETL request process,and makes it easier for analysts to access and navigate[③].

In a snowflake schema, dimensions are not clearly defined. They are normalized to help reduce data redundancy[④] and protect data integrity[⑤]. It takes less space to store dimension tables, but it is a more complicated structure (multiple tables to populate and synchronize) that can be difficult to maintain.

4. Advantages of a Data Mart

Managing big data to gain valuable business insights—is a challenge all companies face, and one that most are answering with strategic data marts.

• Efficient access—A data mart is a time-saving solution for accessing a specific set of data for business intelligence.

• Inexpensive data warehouse alternative[⑥]—Data marts can be an inexpensive alternative to developing an enterprise data warehouse, where required data sets are smaller. An independent data mart can be up and running in a week or less.

• Improving data warehouse performance—Dependent and hybrid data marts can improve the performance of a data warehouse to meet the needs of the analyst. When dependent data marts are placed in a separate processing facility, they significantly reduce analytics processing costs as well.

Other advantages of a data mart include:

• Data maintenance—Different departments can own and control their data.

• Simple setup—The simple design requires less technical skill to set up.

• Analytics—Key performance indicators (KPIs[⑦]) can be easily tracked.

---

① bottom-up method：自底向上法

② blueprint ['bluːprɪnt]　n. 蓝图，设想

③ navigate ['nævɪgeɪt]　v. 导航

④ redundancy [rɪ'dʌndənsi]　n. 冗余

⑤ data integrity：数据完整性

⑥ alternative [ɔːl'tɜːnətɪv]　n. 可供选择的事物　adj. 替代的，备选的

⑦ KPI：关键绩效指标

• Easy entry—Data marts can be the building blocks of a future enterprise data warehouse project.

5. The Future of Data Marts

Even with the improved flexibility and efficiency that data marts offer, big data is still becoming too big for many on-premises solutions. As data warehouses and data lakes move to the cloud, so do data marts.

With a shared cloud-based platform to create and house data, access and analytics become much more efficient. Transient data clusters can be created for short-term analysis, or long-lived clusters can come together for more sustained[①] work. Modern technologies are also separating data storage from compute[②], allowing for ultimate scalability for querying data.

Other advantages of cloud-based dependent and hybrid data marts include:

• Flexible architecture with cloud-native applications;
• Single depository[③] containing all data marts;
• Resources consumed on-demand;
• Immediate real-time access to information;
• Increased efficiency;
• Consolidation[④] of resources that lowers costs;
• Real-time, interactive analytics.

---

① sustained [səs'teɪnd]   adj. 持久的，持续的

② compute [kəm'pjuːt]   n. 计算

③ depository [dɪ'pɒzɪtəri]   n. 存储库，存放处

④ consolidation [kən,sɒlɪ'deɪʃən]   n. 巩固；合并

Unit 3

Bigdata Text A

Data Collection

Data collection is a systematic process of gathering observations or measurements. Whether you are performing research for business, governmental or academic purposes, data collection allows you to gain first-hand knowledge and original insights into your research problem.

While methods and aims may differ between fields, the overall process of data collection remains largely the same.

To collect high-quality data that is relevant to your purposes, follow these four steps.

1. Define the Aim of Your Research

Before you start the process of data collection, you need to identify exactly what you want to achieve. You can start by writing a problem statement: what is the practical or scientific issue that you want to address and why does it matter?

Next, formulate one or more research questions that precisely define what you want to find out. Depending on your research questions, you might need to collect quantitative or qualitative data: Quantitative data is expressed in numbers and graphs and is analyzed through statistical methods. Qualitative data is expressed in words and analyzed through interpretations and categorizations.

If your aim is to test a hypothesis, measure something precisely, or gain large-scale statistical insights, collect quantitative data. If your aim is to explore ideas, understand experiences, or gain detailed insights into a specific context, collect qualitative data. If you have several aims, you can use a mixed methods approach to collects both types of data.

Examples of quantitative and qualitative research aims: You are researching employee perceptions of their direct managers in a large organization.

• The first aim is to assess whether there are significant differences in perceptions of managers across different departments and office locations.

• The second aim is to gather meaningful feedback from employees to explore new ideas for how managers can improve.

You decide to use a mixed-methods approach to collect both quantitative and qualitative data.

2．Choose Your Data Collection Method

Based on the data you want to collect, decide which method is best suited for your research.

Experimental research is primarily a quantitative method. Interviews/focus groups are qualitative methods. Surveys, observations, archival research and secondary data collection can be quantitative or qualitative methods.

Carefully consider what method you will use to gather data that helps you directly answer your research questions (see Table 3-1).

Table 3-1　**Data Collection Methods**

Method	When to use	How to collect data
Experiment	To test a causal relationship	Manipulate variables and measure their effects on others
Survey	To understand the general characteristics or opinions of a group of people	Distribute a list of questions to a sample online, in person or over-the-phone
Interview/ Focus group	To gain an in-depth understanding of perceptions or opinions on a topic	Verbally ask participants open-ended questions in individual interviews or focus group discussions
Observation	To understand something in its natural setting	Measure or survey a sample without trying to affect them
Archival research	To understand current or historical events, conditions or practices	Access manuscripts, documents or records from libraries, depositories or the internet
Secondary data collection	To analyze data from populations that you can't access first-hand	Find existing datasets that have already been collected, from sources such as government agencies or research organizations

3．Plan Your Data Collection Procedures

When you know which method(s) you are using, you need to plan exactly how you will implement them and what procedures you will follow to make accurate observations or measurements of the variables you are interested in.

For instance, if you're conducting surveys or interviews, decide what form the questions will take; if you're conducting an experiment, make decisions about your experimental design.

3.1　Operationalization

Sometimes your variables can be measured directly. For example, you can collect data on the average age of employees simply by asking for dates of birth. However, often you'll be interested in collecting data on more abstract concepts or variables that can't be directly observed.

Operationalization means turning abstract conceptual ideas into measurable observations. When planning how you will collect data, you need to translate the conceptual definition of what you want to study into the operational definition of what you will actually measure.

Example of operationalization: You have decided to use surveys to collect quantitative data. The concept you want to measure is the leadership of managers. You operationalize this concept in two ways:

• Ask managers to rate their own leadership skills on 5-point scales assessing the ability to delegate, decisiveness and dependability.

• Ask their direct employees to provide anonymous feedback on the managers regarding the same topics.

Using multiple ratings of a single concept can help you cross-check your data and assess the test validity of your measures.

3.2　Sampling

You may need to develop a sampling plan to obtain data systematically. This involves defining a population, the group you want to draw conclusions about, and a sample, the group you will actually collect data from.

Your sampling method will determine how you recruit participants or obtain measurements for your study. To decide on a sampling method you will need to consider factors like the required sample size accessibility of the sample and time frame of the data collection.

3.3　Standardizing procedures

If multiple researchers are involved, write a detailed manual to standardize data collection procedures in your study.

This means laying out specific step-by-step instructions so that everyone in your research team collects data in a consistent way—for example, by conducting experiments under the same conditions and using objective criteria to record and categorize observations.

This helps ensure the reliability of your data, and you can also use it to replicate the study in the future.

3.4　Creating a data management plan

Before beginning data collection, you should also decide how you will organize and store your data.

If you are collecting data from people, you will likely need to anonymize and safeguard the data to prevent leaks of sensitive information (e.g. names or identity numbers).

If you are collecting data via interviews or pencil-and-paper formats, you will need to perform transcriptions or data entry in systematic ways to minimize distortion.

You can prevent loss of data by having an organization system that is routinely backed up.

4．Collect the Data

Finally, you can implement your chosen methods to measure or observe the variables you are interested in.

Examples of collecting qualitative and quantitative data: To collect data about perceptions of managers, you administer a survey with closed-and-open-ended questions to a sample of 300 company employees across different departments and locations.

The closed-ended questions ask participants to rate their manager's leadership skills on scales from 1-5. The data produced is numerical and can be statistically analyzed for averages and patterns.

The open-ended questions ask participants for examples of what the manager is doing well now and what they can do better in the future. The data produced is qualitative and can be categorized through content analysis for further insights.

To ensure that high quality data is recorded in a systematic way, here are some best practices:

• Record all relevant information as and when you obtain data. For example, note down whether or how lab equipment is recalibrated during an experimental study.

• Double-check manual data entry for errors.

• If you collect quantitative data, you can assess the reliability and validity to get an indication of your data quality.

New Words

collection	[kə'lekʃn]	n. 收集，采集
systematic	[ˌsɪstə'mætɪk]	adj. 有系统的，有规则的；有步骤的
observation	[ˌɒbzə'veɪʃn]	n. 观测，观察
measurement	['meʒəmənt]	n. 量度；测量
identify	[aɪ'dentɪfaɪ]	v. 识别，认出；确定
formulate	['fɔːmjuleɪt]	v. 构想出，规划；确切地阐述；用公式表示
quantitative	['kwɒntɪtətɪv]	adj. 定量的
qualitative	['kwɒlɪtətɪv]	adj. 定性的
interpretation	[ɪnˌtɜːprɪ'teɪʃn]	n. 解释，说明
categorization	[ˌkætəgəraɪ'zeɪʃn]	n. 分类
hypothesis	[haɪ'pɒθəsɪs]	n. 假设，假说
context	['kɒntekst]	n. 情景；上下文
mix	[mɪks]	v. (使)混合
perception	[pə'sepʃn]	n. 看法；觉察(力)；观念
assess	[ə'ses]	v. 评估；估价；估算
experimental	[ɪkˌsperɪ'mentl]	adj. 根据实验的，试验性的

survey	[ˈsɜːveɪ]	n. 调查
		v. 进行民意测验
causal	[ˈkɔːzl]	adj. 具有因果关系的
manipulate	[məˈnɪpjuleɪt]	v. 操作，处理
distribute	[dɪˈstrɪbjuːt]	v. 发布；分发；分配
verbally	[ˈvɜːbəli]	adv. 言词上；口头地
condition	[kənˈdɪʃn]	n. 状况；环境，处境；条件
conduct	[kənˈdʌkt]	v. 进行，组织，实施
operationalization	[ˌɒpəˈreɪʃnəlaɪˈzeɪʃn]	n. 操作化
abstract	[ˈæbstrækt]	adj. 抽象的，纯理论的
concept	[ˈkɒnsept]	n. 概念；观念
skill	[skɪl]	n. 技能；本领
delegate	[ˈdelɪgeɪt]	v. 授权；委托，委派
decisiveness	[dɪˈsaɪsɪvnəs]	n. 果断性
dependability	[dɪˌpendəˈbɪləti]	n. 可信任，可靠性
anonymous	[əˈnɒnɪməs]	adj. 匿名的，不知名的；不记名的
regard	[rɪˈgɑːd]	v. 认为，看待
cross-check	[krɒs tʃek]	v. 交叉检查；再确认，再次复核
sampling	[ˈsɑːmplɪŋ]	n. 抽样，取样，采样
population	[ˌpɒpjuˈleɪʃn]	n. 总体；人口；族群
participant	[pɑːˈtɪsɪpənt]	n. 参加者，参与者
obtain	[əbˈteɪn]	v. 得到
standardize	[ˈstændədaɪz]	v. 使标准化
instruction	[ɪnˈstrʌkʃn]	n. 指示，指令
replicate	[ˈreplɪkeɪt]	v. 复制；重复，反复
transcription	[trænˈskrɪpʃn]	n. 抄写；录音；翻译
distortion	[dɪˈstɔːʃn]	n. 歪曲，曲解；失真
record	[ˈrekɔːd]	n.&v. 记录，记载
relevant	[ˈreləvənt]	adj. 相关的；合适的；有意义的
recalibrate	[rɪˈkælɪbreɪt]	n. 重新校准
manual	[ˈmænjuəl]	adj. 用手的；手工的；手动的
validity	[vəˈlɪdəti]	n. 有效性，合法性
indication	[ˌɪndɪˈkeɪʃn]	n. 指示，表明

Phrases

| first-hand knowledge | 第一手知识 |
| problem statement | 问题陈述 |

be expressed in	表示为
data collection method	数据收集方法
be suited for …	适合做……，适合于……
focus group	焦点小组，小组讨论
be interested in …	对……有兴趣;
data management plan	数据管理计划
sensitive information	敏感信息
identity number	身份号
data entry	数据输入

 ## Analysis of Difficult Sentences

[1] While methods and aims may differ between fields, the overall process of data collection remains largely the same.

本句中，While methods and aims may differ between fields 是让步状语从句，修饰谓语 remains largely the same。

[2] Carefully consider what method you will use to gather data that helps you directly answer your research questions.

本句中，what method you will use to gather data that helps you directly answer your research questions 作 consider 的宾语。that helps you directly answer your research questions 是定语从句，修饰和限定 data。

[3] When you know which method(s) you are using, you need to plan exactly how you will implement them and what procedures will you follow to make accurate observations or measurements of the variables you are interested in.

本句中，how you will implement them and what procedures you will follow to make accurate observations or measurements of the variables you are interested in 作 plan 的宾语。to make accurate observations or measurements of the variables you are interested in 是动词不定式短语作目的状语。

[4] You can prevent loss of data by having an organization system that is routinely backed up.

本句中，by having an organization system that is routinely backed up 是介词短语，作方式状语，修饰谓语 can prevent。在该短语中，that is routinely backed up 是定语从句，修饰和限定 an organization system。

参考译文

数 据 收 集

数据收集是收集观察或测量的系统过程。无论是出于商业、政府或学术目的进行研究，数据收集都可以让你获得第一手知识和对所研究问题的独到见解。

虽然不同领域数据收集的方法和目标可能不同，但其整体过程基本相同。

要收集与你的目的相关的高质量数据，可按照以下四个步骤操作。

1. 明确研究目标

在开始数据收集过程之前，需要准确确定你要实现的目标。可以先写一个问题陈述：你想解决的实际或科学问题是什么，为什么它很重要？

接下来，确定一个或多个研究问题，精确定义想要了解的内容。根据你所研究的问题，可能需要收集定量或定性数据：定量数据以数字和图表表示，并通过统计方法进行分析。定性数据以文字表达，并通过解释和分类进行分析。

如果你的目标是检验假设、精确度量某物或获得大规模统计见解，可收集定量数据。如果你的目标是探索想法、理解经验或获得对特定环境的详细见解，可收集定性数据。如果有多个目标，则可以使用混合方法来收集这两种类型的数据。

定量和定性研究目标的例子：你正在研究大型组织中员工对其直属经理的看法。

- 首要目标是评估不同部门和办公地点的员工对经理的看法是否存在显著差异。
- 第二个目标是从员工那里收集有意义的反馈，以探索新的管理思维。

你可以使用混合方法来收集定量和定性数据。

2. 选择数据收集方法

根据要收集的数据，决定哪种方法最适合你的研究。

实验研究主要是一种定量方法。访谈/焦点小组是定性方法。调查、观察、档案研究和二手数据收集可以是定量方法，也可以是定性方法。

仔细考虑你将使用什么方法来收集有助于直接回答你所研究的问题的数据(见表 3-1)。

表 3-1　数据收集方法

方　法	何时使用	如何收集数据
实验	检验因果关系	处理变量并度量它们对其他因素的影响
调查	了解人群的总体特征或意见	线上、亲自或通过电话将问题列表分发给抽选人员
访谈/焦点小组	深入了解对某个主题的看法或意见	在个人访谈或焦点小组讨论中口头询问参与者开放式问题
观察	了解自然环境中的事物	评估或调查样本而不试图影响它们
档案研究	了解当前或历史事件、条件或做法	从图书馆、资料库或互联网获取手稿、文件或记录
二手数据收集	分析你无法直接访问的人群数据	从政府机构或研究组织等来源处查找已收集的现有数据集

3. 计划数据收集程序

当你知道要使用哪种方法时，就需要准确地计划如何实施以及将遵循哪些程序来对你感兴趣的变量进行准确的观察或度量。

例如，如果你要进行调查或采访，应决定采取何种形式；如果你正在进行一项实验，应决定你的实验设计。

3.1 操作

有时变量可以直接度量。例如，你可以通过询问出生日期来收集员工平均年龄的数据。但是，你通常会对收集无法直接观察的更抽象概念或变量数据感兴趣。

操作化意味着将抽象的概念想法转化为可衡量的观察结果。在规划如何收集数据时，需要将要研究的概念定义转化为实际度量的操作定义。

操作示例：你已决定使用调查来收集定量数据。你要衡量的概念是管理者的领导力。你可以通过以下两种方式实施此概念：

- 要求管理者以 5 分制评估自己的领导技能，评估授权能力、果断性和可靠性。
- 要求他们的直属员工就相同主题提供对管理者的匿名反馈。

使用单个概念的多个评级可以帮助你交叉检查数据并评估测试的有效性。

3.2 抽样

你可能需要制订一个抽样计划来系统地获取数据。这涉及定义研究总体(你想要得出结论的群体)和样本(你实际从中收集数据的群体)。

你的抽样方法将决定你如何招募参与者或为你的研究获取度量数据。要决定抽样方法，你需要考虑所需的样本量、样本的可访问性和数据收集的时间范围等因素。

3.3 标准化程序

如果涉及多个研究人员，应编写详细的手册以规范你研究中的数据收集程序。

这意味着要制定具体的分步说明，以便研究团队中的每个人都以一致的方式收集数据——例如，通过在相同条件下进行实验并使用客观标准来记录和分类观察结果。

这有助于确保数据的可靠性，未来还可以重复研究。

3.4 制订数据管理计划

在开始数据收集之前，你还应该决定如何组织和存储数据。

如果你从他人那里收集数据，可能需要对数据进行匿名处理并保护数据，以防止泄露敏感信息(例如姓名或身份证号码)。

如果你通过访谈或纸笔的形式收集数据，则需要以系统的方式进行转录或数据输入，以最大限度地减少失真。

你可以定期备份组织系统来防止数据丢失。

4. 收集数据

最后，你可以采用你选择的方法来评估或观察你感兴趣的变量。

收集定性和定量数据的示例：为了收集有关管理者看法的数据，对来自不同部门和地点的 300 名公司员工进行了一项调查，其中包括封闭式和开放式问题。

封闭式问题要求参与者按照 1~5 的等级对管理者的领导技能进行评分。这种方式产生的数据是数字的，可以进行统计分析以得到平均值和模式。

开放式问题要求参与者举例说明管理者目前做得好的地方以及他们将来可以做得更好的地方。这种方式产生的数据是定性的，可以通过内容分析进行分类以获得进一步的见解。

为确保以系统的方式记录高质量的数据，采用以下做法效果更好：

• 在获得数据时记录所有相关信息。例如，记下在实验研究期间是否或如何重新校准实验室设备。

• 仔细检查手动输入数据是否有错误。

• 如果收集定量数据，可以评估信度和效度以了解数据质量。

Bigdata **Text B**

5 Reasons Why You Should Store Big Data in the Cloud

Gone are the days when storage of information can only be done with the traditional remote servers. Today, the in-thing is cloud data storage where information and data are stored electronically online. With this approach, you can store unlimited data online (in the cloud) and access it anywhere. Many articles have been written on storage clouds and benefits of the cloud, but this piece puts forward five of the biggest benefits that you get to enjoy when you save your big data in the cloud.

1．**Easy Accessibility and Usability**

You do not need to be tech-savvy to be able to save your data online. StudyMoose notes that all you have to do in most cases is to drag your file or data and drop it into the online space. It is as easy as that. When you want to access it, what you need is your username and password. Anywhere there is Internet service and an appropriate device, accessing your files should never be an issue. In fact, you don't have to use your devices anytime you plan to access it. A friend's device or a public computer is enough.

2．**Easy Recovery of Data**

Every file is susceptible to a number of disasters and setbacks. When this happens, it could mean the loss of an important business proposal or even a Ph.D. thesis. Cloud storage is the safest means of protecting a file for a long time. Writers of free essays online have learned to save their essay databases in cloud to guard against attack by malware and viruses. Anyone can find free essays from StudyMoose.

So, for your big files that you don't want to lose permanently, a convenient and safe way to save them is to push them to the cloud. They get stored in remote locations from where you can access and retrieve them at any point in time. Similarly, this kind of storage usually has a backup server from which recovery can be made should there be any data loss.

3．Cost-Effectiveness

Storing big data off the cloud can be very expensive and space-consuming. Imagine if you have to save a whole country's database in a disk or hard drive, it would consume a lot of hardware which, in effect, would take up a lot of usable space. Also, it could mean that some resources and power would be expended while trying to save the data. However, with cloud storage, a substantial amount of cost is saved. In fact, storing data digitally online is almost free. The only time you may have to pay is when you want to access it via the internet. No maintenance service is required for big data stored in the cloud, and this is another reason why it is considered cost-effective (see Figure 3-1).

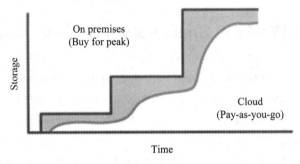

Figure 3-1　Cost-effectiveness

4．Automation and Easy Sharing of Data

Since it is a huge file, you would want some degrees of automation, especially for regular backup. This would ensure that you won't have to keep vigil while trying to ensure that everything is alright. Most free English essays online and other data by big businesses are safeguarded via this means. The essays are safely protected and backed up automatically without having to log in online each time the update is due. And what's more? The process is free of charge. Similarly, if there is a need to share your file or data with friends, clients, colleagues, or anyone, it is quite easy because it is in the cloud. There is no need trying to move files or disk around. All that is needed is to enable the intended recipient to have access to the exact cloud environment where the file is or share the account information with them.

5．Synchronization and Scalable Service

With cloud data storage, you can easily scale resources and pay for only the things you need. This method also allows you to define what the attributes of the cloud would be. The implication of this is that things become more flexible with you, and you can effectively secure your big data. Furthermore, part of the benefits of the cloud storage option for big data is the capacity to synchronize data on your devices. In other words, your data saved in the cloud can be accessed either on your tablet or PC. And if you want to connect the two devices together to access this data, it is equally easy. What this means is that cloud information can be seamlessly transferred among devices without running the risk of complications often associated with a manual exchange of data and files between devices.

Cloud data storage, especially for big data, hardly has any negative effect. Rather, the benefits are very numerous. Beginning from easy access to security, collaboration, file sharing with other users to easy synchronization, the benefits of cloud storage make a good case for it. Good enough, you do not need to have much technical know-how to store your big data in the cloud. For the most part, you only need to know how to log in and out of a platform and save files.

New Words

electronically	[ɪˌlek'trɒnɪkli]	adv. 电子地
accessibility	[əkˌsesə'bɪləti]	n. 可访问性
tech-savvy	[tek 'sævɪ]	adj. 懂技术的
username	['juːzəneɪm]	n. 用户名
password	['pɑːswɜːd]	n. 口令；密码
appropriate	[ə'prəuprɪət]	adj. 适当的，合适的
device	[dɪ'vaɪs]	n. 设备，装置
recovery	[rɪ'kʌvəri]	n. 恢复；改善；失而复得
susceptible	[sə'septəbl]	adj. 易受影响的；易受感染的
setback	['setbæk]	n. 挫折；阻碍
proposal	[prə'pəuzl]	n. 提议，建议
essay	['eseɪ]	n. 短文，文章；论文
virus	['vaɪrəs]	n. 病毒
permanently	['pɜːmənəntli]	adv. 永久地，长期不变地
convenient	[kən'viːnɪənt]	adj. 方便的；实用的
cost-effectiveness	[kɒst ɪˌfek'tɪvnɪs]	n. 成本效用
expend	[ɪk'spend]	v. 花费；消耗
substantial	[səb'stænʃl]	adj. 大量的；结实的，牢固的
ensure	[ɪn'ʃuə]	v. 确保；担保
colleague	['kɒliːg]	n. 同事；同行
synchronization	[ˌsɪŋkrənaɪ'zeɪʃn]	n. 同步；同时性
attribute	[ə'trɪbjuːt]	n. 属性，性质；特征
implication	[ˌɪmplɪ'keɪʃn]	n. 含义；可能的影响(作用、结果)
flexible	['fleksəbl]	adj. 柔韧的；灵活的
connect	[kə'nekt]	v. (使)连接，接通
seamlessly	['siːmləsli]	adv. 无缝地，无空隙地

transfer	[træns'fɜː]	v. 传输
complication	[ˌkɒmplɪ'keɪʃn]	n. 使复杂化的难题(困难)
collaboration	[kəˌlæbə'reɪʃn]	n. 协作，合作
technical	['teknɪkl]	adj. 技术的，技能的
know-how	[nəʊ haʊ]	n. 诀窍；专门技能；专有技术

Phrases

remote server	远程服务器
put forward	提出
drop into	拖入
backup server	备份服务器
hard drive	硬盘驱动器
take up	占用，占据
maintenance service	维修业务，技术维护
be associated with ...	与……有关
negative effect	负面影响
file sharing	文件共享

Abbreviations

PC (Personal Computer)　　　　　　个人计算机

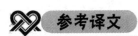 参考译文

应将大数据存储在云中的 5 个原因

只能使用传统远程服务器来存储信息的日子已经一去不复返了。今天，流行的是云数据存储，其中信息和数据以电子方式在线存储。通过这种方法，你可以在线(在云中)存储无限量的数据并能在任何地方访问。许多文章都写了关于存储云和云的益处，本文介绍将大数据保存在云中时你可以享受的五大益处。

1. 易访问性和可用性

你无须精通技术即可在线保存数据。StudyMoose 指出，在大多数情况下，你所要做的就是将文件或数据拖放到在线空间中。就是这么简单。当你想访问时，只需要你的用户名和密码。在任何有互联网服务和适当设备的地方，访问文件都不是问题。事实上，你也不

必总是用自己的设备来访问。朋友的设备或公用电脑就足够了。

2. 轻松恢复数据

　　每个文件都容易受到许多灾难和挫折的影响。发生这种情况时，可能意味着丢失重要的商业提案甚至博士学位论文。云存储是长期保护文件的最安全方法。在线免费文章的作者已经学会了将他们的文章数据库保存在云中，以防受到恶意软件和病毒的攻击。任何人都可以从 StudyMoose 找到免费文章。

　　因此，对于你不想永久丢失的大文件，一种方便且安全的保存方式是将它们推送到云端。它们存储在远程位置，你可以随时访问和检索。同样，这种存储通常有一个备份服务器，如果有任何数据丢失，都可以从中恢复。

3. 成本效益

　　不在云端存储大数据可能非常昂贵且占用空间。想象一下，如果你必须将整个国家的数据库保存在磁盘或硬盘驱动器中，就会消耗大量硬件，实际上就会占用大量可用空间。此外，这可能意味着在保存数据时会消耗一些资源和电力。但是，使用云存储可以节省大量成本。事实上，以数字方式在线存储数据几乎是免费的。只有当你想通过互联网访问时才要付费。存储在云中的大数据不需要维护服务，这也是它被认为具有成本效益的另一个原因(见图3-1)。

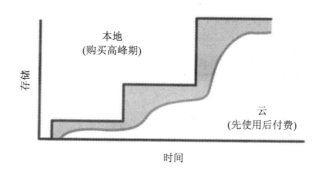

图 3-1　成本效用

4. 数据的自动化和轻松共享

　　由于数据是一个巨大的文件，因此需要一定程度的自动化，特别是定期备份。这将确保你一切正常而不必时时警惕。大多数免费的在线英文文章和大企业的其他数据都通过这种方式得到保护。文章受到安全保护并自动备份，无须每次更新时都在线登录。此外，该过程是免费的。同样，如果需要与朋友、客户、同事或任何人共享你的文件或数据，也都很容易，因为它们存储在云中。无须尝试移动文件或磁盘。需要做的就是让目标接收者能够访问文件所在的确切云环境或与他们共享账户信息。

5. 同步和可扩展的服务

借助云数据存储，你可以轻松扩展资源并只为你需要的东西付费。此方法还允许你定义云的属性。这意味着事情变得更加灵活，你可以有效地保护你的大数据。此外，选择大数据云存储的部分好处是能够在你的设备上同步数据。换句话说，你可以在平板电脑或 PC 上访问你保存在云中的数据。如果你想将两个设备连接在一起以访问这些数据，也同样容易。这意味着云信息可以在设备之间无缝传输，而不必承担在设备之间手动交换数据和文件所带来的复杂风险。

云数据存储，尤其是大数据存储，几乎没有任何负面影响。相反，其好处非常多，如轻松访问、安全性、协作、与其他用户共享文件以及轻松同步文件。你无须具备太多技术知识即可将大数据存储在云中。大多数情况下，你只需要知道如何登录和退出平台并保存文件即可。

Bigdata **Exercises**

〖Ex. 1〗 根据 Text A 回答以下问题。

1. What is data collection?

2. What are the four steps to collect high-quality data that is relevant to your purposes?

3. What do you need to do before you start the process of data collection?

4. When should you collect quantitative data?

5. What does operationalization mean?

6. What can using multiple ratings of a single concept help you?

7. What will your sampling method do?

8. What should you also do before beginning data collection?

9. How can you prevent loss of data?

10. What are some best practices to ensure that high quality data is recorded in a systematic way?

〖Ex. 2〗 根据 Text B 回答以下问题。

1. What is the in-thing today?

2. What are five of the biggest benefits that you get to enjoy when you save your big data in the cloud?

3. What does StudyMoose note?

4. What have writers of free essays online learned to do?

5. What can storing big data off the cloud be?

6. When is the time you may have to pay?

7. Why is it quite easy when you need to share your file or data with friends, clients, colleagues, or anyone?

8. What can you do with cloud data storage?

9. What is part of the benefits of the cloud storage option?

10. What is good enough about cloud data storage?

〖Ex. 3〗 把下列单词或词组翻译成中文。

1. data collection method	1.	
2. data entry	2.	
3. sensitive information	3.	
4. be associate with ...	4.	
5. backup server	5.	
6. hard drive	6.	
7. negative effect	7.	
8. abstract	8.	
9. accessibility	9.	
10. assess	10.	

〖Ex. 4〗 把下列单词翻译成英文。

1. n. 可信任，可靠性	1.
2. v. 发布；分发；分配	2.
3. n. 指示，指令	3.
4. v. 识别，认出；确定	4.
5. adj. 定性的	5.
6. adj. 定量的	6.
7. v. 复制；重复，反复	7.
8. n. 抽样，取样，采样	8.
9. n. 分类	9.
10. n. 收集，采集	10.

〖Ex. 5〗 翻译句子。

1. The first stage in research is data collection.

2. We need to identify actual and potential problems.

3. There is no difference between the two in quantitative terms.

4. There are qualitative differences between the two products.

5. The equipment is still at the experimental stage.

6. Data can be manipulated to enhance image quality.

7. Each programming language has its own means of manipulating data.

8. Ten years on, the original concept was wearing well.

9. If he improved his IT skills, he'd easily get a job.

10. If you do store sensitive information, use windows security to help secure it.

Bigdata **Reading Material**

Big Data Backup

Big data occupies[①] a massive data storage capacity. A backup process for your big data will consume a lot of storage and network resources and it is not an easy task. This raises the question whether it is worthwhile to backup your big data.

1. The Importance of Backup for Big Data

Big data applications were traditionally designed to analyze past product sales and forecast trends[②] and were not considered business-critical. Big data analysis has evolved to be an essential[③] tool for business daily operation. Many customer service applications are using big data analysis to interact with their customers.

Big data management is dependent on the reports generated by big-data-based business analytics tools for daily decisions. These reports can help in making decisions to improve services, product turnaround[④] and profits. Companies invest a lot in big data storage and software. In order to protect this investment, you should regularly back up your big data storage.

2. Backup in the Cloud vs. On-Premises

You can use on-premise storage or a cloud storage service, or both, for big data backup.

2.1 On-premises big data backup

Big data backup requires high volumes of data to be processed at a high rate of input/output operations per second (IOPS[⑤]). On-premises storage devices can be selected to meet these requirements. If a high IOPS rate is required, an on-premise storage device is recommended.

---

① occupy ['ɒkjupaɪ] v. 占用

② forecast trend：预报趋势

③ essential [ɪ'senʃl] adj. 基本的，必不可少的

④ product turnaround：产品周转

⑤ IOPS：每秒的输入/输出操作

2.2 Big data backup in the cloud

Cloud-based services are very versatile[①] and can be a solution for most of your big data backup needs. Many companies are using cloud services for big data storage and processing. It would be efficient to also backup big data over the cloud. There are several advantages to using a cloud service. Cloud backup can be replicated[②] in multiple locations for redundancy. The data you back up in the cloud is secured with advanced encryption techniques.

2.3 On-premises and cloud hybrid solution

You can use both on-premises infrastructure[③] and cloud services for your big data backup. In this way, you can efficiently use all these resources to make your backup process cost-effective. Having your data backed up on-premises and in the cloud can add another layer of protection.

Using both solutions can also be helpful for disaster[④] recovery. In the case of on-site disaster, cloud data backups ensure that data remains available. But it may also make sense to keep your backups on-site. In cases where the disaster did not damage the backup volumes, the data will be restored more quickly compared to the remote sites.

3. Best Practices for Big Data Backup

Here are several tips on how to back up your big data.

• Backup in a remote location—when protecting your big data from physical disasters, such as fires, earthquakes, and storms, it is important that your backup resides in off-site and remote locations.

• Data backup is not enough for disaster recovery—you should remember there is a difference between data backup and disaster recovery. Normally, you restore from your backup a particular file or directory which was corrupted or overwritten. But in a case of disaster recovery, you are going to recover all your data. This may take a long time. When a disaster occurs, you need a disaster recovery plan to ensure your big data operations return to normal as soon as possible.

• Using snapshots[⑤]—snapshots are used as a backup copy to protect against data loss. Snapshots can be used for big data backup only when the data is not changing rapidly. If you are storing your big data in Amazon cloud, you can use the AWS snapshots service.

• Data compression—you can decrease your big data size by compressing[⑥] it. This will

---

① versatile ['vɜːsətaɪl]   adj. 多用途的，多功能的
② replicate ['replɪkeɪt]   vt. 复制
③ infrastructure ['ɪnfrəstrʌktʃə]   n. 基础设施
④ disaster [dɪ'zɑːstə]   n. 灾难
⑤ snapshot ['snæpʃɒt]   n. 快照
⑥ compress [kəm'pres]   v. (使)压缩

result in reducing storage costs and bandwidth[1] usage. For data compression, you can use data deduplication[2]. Data deduplication is a process that eliminates redundant data blocks within a dataset.

• Schedule frequent backups—to prevent loss of data, you should know how often your data is changed. According to this, you should set the frequency of your backups. You can also create a different backup schedule for different data blocks[3]. For example, you can set your backup application that specific data blocks are backed up once a day, while others are backed up only once a week. This will eliminate[4] unnecessary backups, reducing storage and costs.

• Backup retention[5]—you should consider the time period you want to keep your backup on the storage server. Keeping a long history of your big data backups will consume a lot of storage space, which eventually will run out[6]. Instead, you can set a more feasible[7] retention policy. An example of such a policy is: hourly and daily backups are kept for a week, weekly backups are kept for a month, and monthly backups for six months.

• Backup security—a good security practice is to have your big data backups protected by strong data encryption technology[8]. This will ensure that in case of unauthorized access to your backup, the data will be unusable. The data should be encrypted while it is at rest, inside the storage device, and while it is on the move from one location to another.

---

① bandwidth ['bændwɪdθ]   n. 带宽

② deduplication [dɪdjuːplɪ'keɪʃn]   n. 去重复

③ data block：数据块

④ eliminate [ɪ'lɪmɪneɪt]   v. 清除，排除，清除

⑤ retention [rɪ'tenʃn]   n. 保留

⑥ run out: 用完；耗尽

⑦ feasible ['fiːzəbl]   adj. 可行的，做得到的

⑧ data encryption technology：数据加密技术

Unit 4

Text A

Data Preprocessing

1. The Meaning of Data Preprocessing

Data preprocessing involves transforming raw data to well-formed data sets so that data mining analytics can be applied. Raw data is often incomplete and has inconsistent formatting. The adequacy or inadequacy of data preparation has a direct correlation with the success of any project that involve data analyics.

Preprocessing involves both data validation and data imputation. The goal of data validation is to assess whether the data in question is both complete and accurate. The goal of data imputation is to correct errors and input missing values—either manually or automatically through business process automation (BPA) programming.

Data preprocessing is used in both database-driven and rules-based applications. In machine learning (ML) processes, data preprocessing is critical for ensuring that large datasets are formatted in such a way that the data they contain can be interpreted and parsed by learning algorithms.

2. Importance of Data Preprocessing

When using data sets to train machine learning models, you'll often hear the phrase "garbage in, garbage out". This means that if you use bad or "dirty" data to train your model, you'll end up with a bad, improperly trained model. And the model won't actually be relevant to your analysis.

Good preprocessed data is even more important than the most powerful algorithms. Machine learning models trained with bad data could actually be harmful to the analysis you're trying to do—giving you "garbage" results.

67

Depending on your data gathering techniques and sources, you may end up with data that's out of range or that includes an incorrect feature, like household income below zero. Your set could have missing values or fields. Text data, for example, will often have misspelled words and irrelevant symbols, URLs, etc.

When you properly preprocess and clean your data, you'll set yourself up for much more accurate downstream processes. We often hear about the importance of "data-driven decision making", but if these decisions are driven by bad data, they're simply bad decisions.

3. Data Quality Assessment

After you have properly gathered the data, you need to explore or assess it to spot key trends and inconsistencies. The main goals of data quality assessment are:

• Get data overview: that is to understand the data formats and overall structure in which the data is stored. It is also to find the properties of data like mean, median, standard quantiles and standard deviation. These details can help identify irregularities in the data.

• Identify missing data: missing data is common in most real-world datasets. It can disrupt true data patterns, and even lead to more data loss when entire rows and columns are removed because of a few missing cells in the dataset.

• Identify outliers or anomalous data: some data points fall far out of the predominant data patterns. These points are outliers. They might need to be discarded to get predictions with higher accuracies unless the primary purpose of the algorithm is to detect anomalies.

• Remove inconsistencies: just like missing values, real-world data also has multiple inconsistencies like incorrect spellings, incorrectly populated columns and rows (eg. salary populated in gender column), duplicated data and much more. Sometimes, these inconsistencies can be treated through automation, but most often they need a manual check-up.

4. Data Preprocessing Steps

4.1 Data cleansing

Data cleansing is the process of altering data in a given storage resource to make sure that it is accurate and correct. Data cleansing is also known as data cleaning or data scrubbing.

Data cleansing is sometimes compared to data purging, which is to delete old or useless data from a data set. Although data cleansing can involve deleting old, incomplete or duplicated data, it is different from data purging. Data purging usually focuses on clearing space for new data, whereas data cleansing focuses on maximizing the accuracy of data in a system. Parsing or other methods may be used to get rid of syntax errors, typographical errors or fragments of records. Careful analysis of a data set can show how merging multiple sets lead to duplication. In this case data cleansing may be used to fix the problem (see Figure 4-1).

Figure 4-1   Data Cleansing Cycle

4.2   Data integration

Data integration is a process in which heterogeneous data is retrieved and combined as an incorporated form and structure. Data integration allows different data types (such as data sets, documents and tables) to be merged by users, organizations and applications as personal or business processes and/or functions.

Data integration primarily supports the analytical processing of large data sets by aligning, combining and presenting each data set from organizational departments and external remote sources to fulfill integrator objectives.

Data integration is generally implemented in data warehouses through specialized software that hosts large data repositories from internal and external resources. Data is extracted, integrated and presented as a unified form. For example, a user's complete data set may include extracted and combined data from marketing, sales and operations. And the data is combined to form a complete report.

An example of data integration in a smaller paradigm is spreadsheet integration in a Microsoft Word document.

4.3   Data transformation

Data is normalized and generalized. Normalization is a process that ensures that no data is redundant, it is all stored in a single place, and all the dependencies are logical.

4.4   Data reduction

When the volume of data is huge, databases can become slower, costly to access, and challenging to properly store. Data reduction aims to present a reduced representation of the data in a data warehouse.

There are various methods to reduce data. For example, once a subset of relevant attributes is chosen for its significance, anything below a given level is discarded.

Encoding mechanisms can be used to reduce the size of data as well. If all original data can be recovered after compression, the operation is labeled as lossless. If some data is lost, then it's called a lossy reduction. Aggregation can also be used to condense countless transactions into a single weekly or monthly value, significantly reducing the number of data objects (see Figure 4-2).

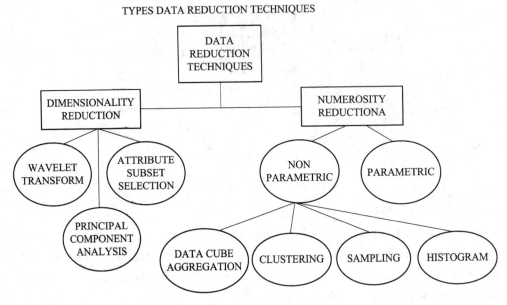

Figure 4-2　Data Reduction Techniqus

4.5　Data deduplication

Data deduplication is a data compression technique in which redundant or repeated copies of data are removed from a system. It is implemented in data backup and network data mechanisms. It enables the storage of one unique instance of data within a database or information system (IS).

Data deduplication is also known as intelligent compression, single instance storage, commonality factoring or data reduction.

Data deduplication works by analyzing and comparing incoming data segments with previously stored data. If data is already present, data deduplication algorithms discard the new data and create a reference. For example, if a document file is backed up with changes, the previous file and applied changes are added to the data segment. However, if there is no difference, the newer data file is discarded, and a reference is created. Similarly, a data deduplication algorithm scans outgoing data on a network connection to check for duplicates, and remove them to increase data transfer speed.

4.6　Data sampling

Sometimes, due to time, storage or memory constraints, a dataset is too big or too complex to be worked with. Sampling techniques can be used to select and work with just a subset of the dataset, provided that it has approximately the same properties of the original one.

New Words

preprocessing	[priːˈprəʊsesɪŋ]	n. 预处理，预加工
inconsistent	[ˌɪnkənˈsɪstənt]	adj. 不一致的，前后矛盾的，不合逻辑的
adequacy	[ˈædɪkwəsi]	n. 足够，充足；适当，恰当
inadequacy	[ɪnˈædɪkwəsi]	n. 不充分，不完全；不适当
preparation	[ˌprepəˈreɪʃn]	n. 准备(工作)
correlation	[ˌkɒrəˈleɪʃn]	n. 关联，联系
imputation	[ˌɪmpjʊˈteɪʃn]	n. 填补
validation	[ˌvælɪˈdeɪʃn]	n. 确认；有效；校验
rules-based	[ruːlz beɪst]	adj. 规则导向的，基于规则的
parse	[pɑːz]	v. 从语法上描述或分析(词句等)
improperly	[ɪmˈprɒpəli]	adv. 不正确地，不适当地
misspell	[ˌmɪsˈspel]	v. 拼错，写错
symbol	[ˈsɪmbl]	n. 符号，记号
downstream	[ˌdaʊnˈstriːm]	adv. 在下游地
		adj. 顺流而下的；在下游方向的
spot	[spɒt]	v. 注意到
inconsistency	[ˌɪnkənˈsɪstənsi]	n. 不一致，不协调；前后矛盾
mean	[miːn]	n. 平均值
median	[ˈmiːdɪən]	adj. 中值的
		n. 中值；中位数
irregularity	[ɪˌregjəˈlærəti]	n. 不合规；不规则；非正式
disrupt	[dɪsˈrʌpt]	v. 扰乱
remove	[rɪˈmuːv]	v. 删除，移开
outlier	[ˈaʊtˌlaɪə]	n. 离群值；异常值
predominant	[prɪˈdɒmɪnənt]	adj. 主要的，占主导地位的
anomaly	[əˈnɒməli]	n. 异常，反常；不规则
check-up	[tʃek ʌp]	n. 核对，检查
typographical	[ˌtaɪpəˈgræfɪkl]	adj. 印刷上的
fragment	[ˈfrægmənt]	n. 分段
fix	[fɪks]	v. 修理；找到；固定，确定
heterogeneous	[ˌhetərəˈdʒiːnɪəs]	adj. 各种各样的；异构的
structure	[ˈstrʌktʃə]	n. 结构，构造
transformation	[ˌtrænsfəˈmeɪʃn]	n. 转换；变化
normalize	[ˈnɔːməlaɪz]	v. 使正常化；使标准化

generalize	['dʒenrəlaɪz]	v. 普通化，一般化
redundant	[rɪ'dʌndənt]	adj. 多余的，冗余的
significance	[sɪg'nɪfɪkəns]	n. 重要性；含义
discard	[dɪ'skɑːd]	v. 丢弃，抛弃
compression	[kəm'preʃən]	n. 压缩
lossless	['lɒsləs]	adj. 无损的
aggregation	[,ægrɪ'geɪʃn]	n. 聚合，聚集；集成
deduplication	[dɪdjuːplɪ'keɪʃn]	n. 去除重复
instance	['ɪnstəns]	n. 实例，例子
commonality	['kɒmənəlti]	n. 共同特征，共性
scan	[skæn]	v. 扫描
approximately	[ə'prɒksɪmətli]	adv. 大约，近似

Phrases

data preprocessing	数据预处理
well-formed data set	有良好格式的数据集
missing value	遗漏值，缺失值
learning algorithm	学习算法
garbage in/garbage out	垃圾进/垃圾出
end up with ...	以……结束
train model	训练模型
data-driven decision making	数据驱动决策
standard quantiles	标准分位数
standard deviation	标准偏差
lead to ...	导致……
anomalous data	异常数据
data cleansing	数据清理
data cleaning	数据清洗
data scrubbing	数据洗刷
data purging	数据清除
syntax error	语法错误
data integration	数据集成，数据整合
data type	数据类型
unified form	统一形式
data reduction	数据精简
encoding mechanism	编码机制

original data	原始数据
lossy reduction	有损精简
data deduplication	数据去重
data compression technique	数据压缩技术
data transfer speed	数据传输速度

Abbreviations

BPA (Business Process Automation)	业务流程自动化
URL (Uniform Resource Locator)	统一资源定位器，统一资源定位符
IS (Information System)	信息系统

 ## Analysis of Difficult Sentences

[1] In machine learning (ML) processes, data preprocessing is critical for ensuring that large datasets are formatted in such a way that the data they contain can be interpreted and parsed by learning algorithms.

本句中，that large datasets are formatted in such a way that the data they contain can be interpreted and parsed by learning algorithms 是宾语从句，作 ensuring 的宾语。在该从语中，that the data they contain can be interpreted and parsed by learning algorithms 是定语从句，修饰和限定 a way。they contain 是定语从句，修饰和限定 the data。by learning algorithms 是介词短语，作方式状语，修饰谓语 can be interpreted and parsed。

[2] Machine learning models trained with bad data could actually be harmful to the analysis you're trying to do—giving you "garbage" results.

本句中，trained with bad data 是过去分词短语，作定语，修饰和限定 Machine learning models。you're trying to do 是定语从句，修饰和限定 the analysis。

[3] They might need to be discarded to get predictions with higher accuracies unless the primary purpose of the algorithm is to detect anomalies.

本句中，unless the primary purpose of the algorithm is to detect anomalies 是条件状语从句，修饰谓语 need to be discarded。unless 等于 if not，意思是"除非"。

[4] Data integration primarily supports the analytical processing of large data sets by aligning, combining and presenting each data set from organizational departments and external remote sources to fulfill integrator objectives.

本句中，by aligning, combining and presenting each data set from organizational departments and external remote sources 是介词短语，作方式状语，修饰谓语 supports。to fulfill integrator objectives 是动词不定式短语，作目的状语，也修饰谓语 supports。

参考译文

数 据 预 处 理

1．数据预处理的意义

数据预处理涉及将原始数据转换为具有良好格式的数据集，以便应用数据挖掘分析。原始数据通常不完整且格式不一致。数据准备得充分与否与涉及数据分析的项目的成功直接相关。

预处理涉及数据验证和数据填补。数据验证的目标是评估相关数据是否完整和准确。数据填补的目标是纠正错误并输入缺失值——通过手动业务流程自动化(BPA)编程自动进行。

数据预处理用于数据库驱动和基于规则的应用程序。在机器学习(ML)过程中，数据预处理对确保大型数据集的格式能够使其包含的数据能够被学习算法解释和解析至关重要。

2．数据预处理的重要性

在使用数据集训练机器学习模型时，经常会听到"垃圾输入，垃圾输出"这句话。这意味着，如果使用坏的或"脏"数据来训练模型，将得到一个坏的、训练不当的模型，而该模型实际上与你的分析无关。

良好的预处理数据甚至比最强大的算法更重要，用不良数据训练的机器学习模型实际上可能对你所尝试进行的分析有害——输出"垃圾"结果。

根据数据收集技术和来源，你最终可能会得到超出范围或包含不正确特征的数据，例如家庭收入低于零。集合可能缺少值或字段，例如文本数据通常会有拼写错误的单词和不相关的符号、URL 等。

当正确预处理和清理数据时，将为更准确的下游流程做好了准备。我们经常听到"数据驱动决策"的重要性，但如果这些决策是由坏的数据驱动的，那么它们就是糟糕的决策。

3．数据质量评估

正确收集数据后，需要对其进行探索或评估，以发现关键趋势和不一致之处。数据质量评估的主要目标是：

• 获取数据概览：即了解数据存储的数据格式和整体结构。还可以找到数据的特性，如均值、中位数、标准分位数和标准差。这些详细信息可以帮助识别数据中的异常行为。

• 识别缺失数据：缺失数据在大多数数据集实例中很常见。它可能会破坏真实的数据模式，甚至在数据集中缺少几个单元格而删除整行和整列时，会导致更多数据丢失。

• 识别异常值或异常数据：一些数据点与主要数据模式相差甚远。这些点是异常值。除非算法的主要目的是检测异常，否则可能需要丢弃这些点以获得更准确的预测。

· 消除不一致：就像缺失值一样，数据也有多种不一致，如拼写不正确、列和行填充不正确(例如，在性别列中填充工资)、重复数据等等。有时，这些不一致可以自动化处理，但大多数情况下需要手动检查。

4．数据预处理步骤

4.1 数据清理

数据清理是更改给定存储资源中的数据以确保准确无误的过程。数据清理也称为数据清洁或数据刷洗。

有时候把数据清理比作数据清除，即把旧的或无用的数据从数据集中删除。虽然数据清理可能涉及删除旧的、不完整或重复的数据，但数据清理与数据清除不同。数据清除通常侧重于为新数据清理空间，而数据清理则侧重于最大限度地提高系统中数据的准确性。可以使用解析或其他方法来消除语法错误、印刷错误或记录片段。仔细分析数据集可以显示合并多个集如何导致重复。在这种情况下，可以使用数据清理来解决问题(见图 4-1)。

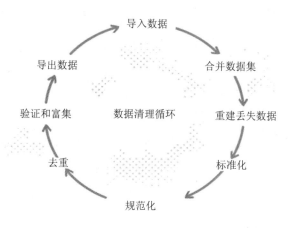

图 4-1　数据清理循环

4.2 数据集成

数据集成是检索异构数据并将其组合为合并形式和结构的过程。数据集成允许用户、组织和应用程序将不同的数据类型(例如数据集、文档和表格)合并为个人或业务流程和/或功能。

数据集成主要通过对齐、组合和呈现来自组织部门和外部远程来源的每个数据集，用于支持大型数据集的分析处理，以实现集成者的目标。

数据集成通常通过专门的软件在数据仓库中实现，该软件托管来自内部和外部资源的大型数据存储库。数据被提取、集成并以统一的形式呈现。例如，一个用户的完整数据集可能包括从市场、销售和运营中提取和组合的数据。这些数据组合在一起形成一个完整的报告。

较小范式数据集成的一个示例是微软 Word 文档中的电子表格集成。

4.3 数据转换

数据被规范化和通用化。规范化是确保没有数据冗余的过程,所有数据都存储在一个地方,并且所有依赖关系都是合乎逻辑的。

4.4 数据精简

当数据量巨大时,数据库会变得更慢、访问成本高并且难以正确存储。数据精简旨在简化数据仓库中的数据。

有多种方法可以精简数据。例如,若按照重要性选择了相关属性的子集,那么低于给定级别的内容都将被丢弃。

也可以用编码机制来减少数据的规模。如果压缩后所有原始数据都可以恢复,则该操作标记为无损。如果某些数据丢失,则称为有损精简。还可以用聚合将大量业务压缩为每周或每月的值,从而显著减少数据对象的数量(见图4-2)。

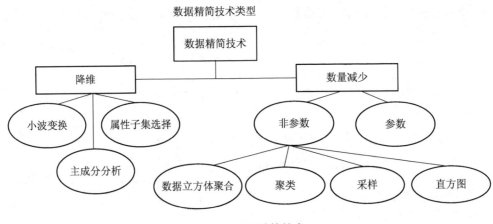

图 4-2 数据精简技术

4.5 数据去重

数据去重是一种数据压缩技术,它从系统中删除冗余或重复的数据副本。数据去重在数据备份和网络数据机制中实施。它能够在数据库或信息系统(IS)内存储一个唯一的数据实例。

数据去重也称为智能压缩、单实例存储、共性分解或数据精简。

数据去重的工作原理是分析输入的数据段并将其与先前存储的数据进行比较。如果数据已经存在,数据去重算法会丢弃新数据并创建引用。例如,如果一个文档文件备份有更改,则先前的文件和应用中的更改将添加到数据段中。但是,如果没有差异,则丢弃较新的数据文件,并创建引用。同理,数据去重算法会扫描网络连接上的输出数据以检查重复项,删除重复项以提高数据传输速度。

4.6 数据采样

有时,由于时间、存储或内存限制,数据集会因太大或太复杂而无法使用。采样技术可用于选择和处理数据集的子集,前提是子集具有与原始数据大致相同的属性。

Bigdata **Text B**

Data Transformation

Analyzing information requires structured and accessible data for best results. Data transformation enables organizations to alter the structure and format of raw data as needed.

1. What Is Data Transformation?

Data transformation is the process of changing the format, structure, or values of data. For data analytics projects, data may be transformed at two stages of the data pipeline. Organizations that use on-premises data warehouses generally use an ETL (extract, transform, load) process, in which data transformation is the middle step. Today, most organizations use cloud-based data warehouses, which can scale compute and storage resources with latency measured in seconds or minutes. The scalability of the cloud platform lets organizations skip preload transformations and load raw data into the data warehouse, then transform it at query time—a model called ELT (extract, load, transform).

Processes such as data integration, data migration, data warehousing and data wrangling all may involve data transformation.

Data transformation may be constructive (adding, copying and replicating data), destructive (deleting fields and records), aesthetic (standardizing salutations or street names), or structural (renaming, moving, and combining columns in a database).

An enterprise can choose among a variety of ETL tools that automate the process of data transformation. Data analysts, data engineers, and data scientists also transform data using scripting languages such as Python or domain-specific languages like SQL.

2. Benefits and Challenges of Data Transformation

Transforming data yields several benefits:

• Data is transformed to make it better-organized. Transformed data may be easier for both humans and computers to use.

• Properly formatted and validated data improves data quality and protects applications from potential landmines such as null values, unexpected duplicates, incorrect indexing, and incompatible formats.

• Data transformation facilitates compatibility between applications, systems, and types of data. Data used for multiple purposes may need to be transformed in different ways.

However, there are challenges to transforming data effectively:

• Data transformation can be expensive. The cost is dependent on the specific infrastructure, software, and tools used to process data. Expenses may include those related to licensing, computing resources, and hiring necessary personnel.

• Data transformation processes can be resource-intensive. Performing transformations in an on-premises data warehouse can create a computational burden that slows down other operations. If you use a cloud-based data warehouse, you can do the transformations after loading because the platform can scale up to meet your demand.

• Lack of expertise and carelessness can introduce problems during transformation. Data analysts without appropriate subject matter expertise are less likely to notice typos or incorrect data because they are less familiar with the range of accurate and permissible values. For example, if someone working on medical data is unfamiliar with relevant terms, he might fail to flag disease names or notice misspellings.

• Enterprises can perform transformations that don't suit their needs. A business might want to change information to a specific format for one application, only to revert the information back to its prior format for a different application.

3. How to Transform Data?

Data transformation can increase the efficiency of analytic and business processes and enable better data-driven decision-making. The first phase of data transformations should include things like data type conversion and flattening of hierarchical data. These operations shape data to increase compatibility with analytics systems. Data analysts and data scientists can implement further transformations additively as needed. Each layer of processing should be designed to perform a specific set of tasks that meet a known business or technical requirement.

Data transformation serves many functions within the data analytics stack.

3.1 Extraction and parsing

In the modern ELT process, data ingestion begins with extracting information from a data source, followed by copying the data to its destination. Initial transformations are focused on shaping the format and structure of data to ensure its compatibility with both the destination system and the data already there. Parsing fields out of comma-delimited log data for loading to a relational database is an example of this type of data transformation.

3.2 Mapping and translation

Some of the most basic data transformations involve the mapping and translation of data. For example, a column containing integers representing error codes can be mapped to the relevant error descriptions, making that column easier to understand and more useful for display in a customer-facing application.

Translation converts data from formats used in one system to formats appropriate for a different system. Even after parsing, web data might arrive in the form of hierarchical JSON or XML files, but need to be translated into row and column data for inclusion in a relational database.

3.3 Filtering, aggregation and summarization

Data transformation is often concerned with whittling data down and making it more manageable. Data may be consolidated by filtering out unnecessary fields, columns, and records. Data might also be aggregated or summarized by, for instance, transforming a time series of customer transactions to hourly or daily sales counts.

BI tools can do this filtering and aggregation, but it can be more efficient to do the transformations before a reporting tool accesses the data.

3.4 Enrichment and imputation

Data from different sources can be merged to create denormalized enriched information. A customer's transactions can be rolled up into a grand total and added into a customer information table for quicker reference or for use by customer analytics systems. Long or freeform fields may be split into multiple columns, and missing values can be imputed or corrupted data replaced as a result of these kinds of transformations.

3.5 Indexing and ordering

Data can be transformed so that it's ordered logically or to suit a data storage scheme. In relational database management systems, for example, creating indexes can improve performance or improve the management of relationships between different tables.

3.6 Anonymization and encryption

Data containing personally identifiable information, or other information that could compromise privacy or security, should be anonymized before propagation. Encryption of private data is a requirement in many industries, and systems can perform encryption at multiple levels, from individual database cells to entire records or fields.

3.7 Modeling, typecasting, formatting and renaming

Finally, a whole set of transformations can reshape data without changing content. This includes converting data types for compatibility, adjusting dates and times, and renaming schemas, tables, and columns for clarity.

3.8 Refining the data transformation process

Before your enterprise can run analytics, and even before you transform the data, you must replicate it to a data warehouse architected for analytics. Most organizations today choose a cloud data warehouse, allowing them to take full advantage of ELT.

New Words

stage	[steɪdʒ]	n. 阶段
latency	['leɪtənsɪ]	n. 延迟
measure	['meʒə]	v. 衡量；测量；量度

preload	[ˌpriːˈləʊd]	v. 预载，预装
constructive	[kənˈstrʌktɪv]	adj. 建设性的；积极的，有助益的
destructive	[dɪˈstrʌktɪv]	adj. 破坏性的；毁灭性的
delete	[dɪˈliːt]	v. 删除
aesthetic	[iːsˈθetɪk]	adj. 审美的；艺术的 n. 美感；审美观
structural	[ˈstrʌktʃərəl]	adj. 结构化的，结构(上)的
rename	[ˌriːˈneɪm]	v. 重命名
combine	[kəmˈbaɪn]	v. 组合，结合，混合
column	[ˈkɒləm]	n. 栏；队
compatibility	[kəmˌpætəˈbɪləti]	n. 兼容性
license	[ˈlaɪsns]	n. 执照，许可证 v. 同意；发许可证
burden	[ˈbɜːdn]	n. 负担；重负 v. (使)担负
expertise	[ˌekspɜːˈtiːz]	n. 专门技术，专门知识
carelessness	[ˈkeələsnəs]	n. 粗心大意；草率
typo	[ˈtaɪpəʊ]	n. 打印错误；打字(排印)文稿的小错误
permissible	[pəˈmɪsəbl]	adj. 许可的；得到准许的
shape	[ʃeɪp]	v. 塑造；准备(做某动作) n. 形状，图形
stack	[stæk]	n. 堆栈
extraction	[ɪkˈstrækʃn]	n. 提取，开采
comma-delimited	[ˈkɒmə dɪˈlɪmɪtɪd]	adj. 用逗号分隔的
map	[mæp]	v. 映射
display	[dɪˈspleɪ]	v. 显示，展示
customer-facing	[kʌstəmə ˈfeɪsɪŋ]	adj. 面对客户的
convert	[kənˈvɜːt]	v. 转变，转换
summarization	[ˌsʌmərɪˈzeɪʃən]	n. 摘要，概要，概括
unnecessary	[ʌnˈnesəsərə]	adj. 不需要的，没必要的
enrichment	[ɪnˈrɪtʃmənt]	n. 富集；浓缩
denormalize	[dɪˈnɔːrməlaɪz]	v. 非规范化，反正规化
enrich	[ɪnˈrɪtʃ]	v. 充实；使丰富
reference	[ˈrefrəns]	v. 参考；引用
freeform	[ˈfriːfɔːrm]	adj. 形式自由的，任意的
corrupt	[kəˈrʌpt]	adj. 损坏的，有错误的

index	[ˈɪndeks]	n. 索引；指数；指示；标志
		v. 给……编索引
order	[ˈɔːdə]	v. 排序
anonymization	[əˌnɒnɪmaɪˈzeɪʃn]	n. 匿名化
compromise	[ˈkɒmprəmaɪz]	v. 危及；妥协；违背
cell	[sel]	n. 单元
typecast	[ˈtaɪpkɑːst]	v. 类型转换
reshape	[ˌriːˈʃeɪp]	v. 重塑；给……以新形态；采取新形式
clarity	[ˈklærəti]	n. 清洗，清楚，明确

Phrases

data pipeline	数据管道，数据流水线
cloud platform	云平台
data wrangling	数据整理
a variety of	各种各样的，多种的
scripting language	脚本语言
null value	空值
be dependent on	依靠，依赖
be designed to	目的是；被设计用于
be consolidated by	合并
be merged to	合并到
split into	(使)分成
data storage scheme	数据存储方案
personally identifiable information	个人身份识别信息
take full advantage	充分利用

Abbreviations

| ELT (Extract, Load, Transform) | 提取、装载、转换 |

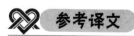 参考译文

数 据 转 换

使用结构化且可访问的数据分析信息才能获得最佳结果。数据转换使组织能够根据需

要来更改原始数据的结构和格式。

1．什么是数据转换？

数据转换是改变数据格式、结构或值的过程。对于数据分析项目，数据可能会在数据管道的两个阶段进行转换。使用本地数据仓库的组织通常使用 ETL(提取、转换、加载)过程，其中数据转换是中间步骤。如今，大多数组织都使用基于云的数据仓库，它能够以秒或分钟为延迟单位来扩展计算和存储资源。云平台的可扩展性使组织可以跳过预加载转换，将原始数据加载到数据仓库中，然后在查询时对其进行转换——这是一种称为 ELT(提取、加载、转换)的模型。

数据集成、数据迁移、数据仓库和数据整理等流程都可能涉及数据转换。

数据转换可能是建设性的(添加、拷贝和复制数据)、破坏性的(删除字段和记录)、审美的(标准化称呼或规范名称)或结构性的(重命名、移动和组合数据库中的列)。

企业可以在多种 ETL 工具中选择，这些工具可以自动执行数据转换过程。数据分析师、数据工程师和数据科学家还使用脚本语言(如 Python)或特定领域的语言(如 SQL)转换数据。

2．数据转换的益处和挑战

转换数据有以下几个益处：
- 数据经过转换，使其组织得更好。转换后的数据可能更易于人和计算机使用。
- 正确格式化和验证的数据可以提高数据质量并保护应用程序免受潜在地雷影响，例如空值、意外重复、不正确的索引和不兼容的格式。
- 数据转换促进了应用程序、系统和数据类型之间的兼容性。用于多种目的的数据可能需要以不同的方式进行转换。

然而，有效地转换数据也存在以下几个挑战：
- 数据转换可能非常昂贵。成本取决于用于处理数据的特定基础设施、软件和工具。可能包括与许可、计算资源和雇用必要人员相关的费用。
- 数据转换过程可能是资源密集型的。在内部部署数据仓库中执行转换可能会造成计算负担，从而减缓其他操作。如果使用基于云的数据仓库，可以在加载后进行转换，因为平台可以扩展从而满足需求。
- 缺乏专业知识和粗心大意可能会在转换过程中造成问题。没有适当专业知识的数据分析师可能不太注意到拼写错误或不正确的数据，因为他们不太熟悉准确和允许值的范围。例如，如果处理医疗数据的人不熟悉相关术语，可能就无法标记疾病名称或不能注意到拼写错误。
- 企业可能执行不适合其需求的转换。企业可能想将信息更改为某个应用程序的特定格式，结果却将信息恢复成另一个应用程序的先前格式。

3．如何转换数据

数据转换可以提高分析和业务流程的效率，并实现更好的数据驱动决策。数据转换的第一阶段应该包括数据类型转换和分层数据的扁平化等内容。这些操作可以塑造数据以提高与分析系统的兼容性。数据分析师和数据科学家可以根据需要额外实施进一步的转换。每一层处理都应设计为执行一组特定的任务以满足已知的业务或技术要求。

数据转换在数据分析堆栈中提供了许多功能。

3.1 提取与解析

在现代 ELT 流程中,数据获取首先从数据源中提取信息,然后将数据复制到其目的地。初始转换侧重于塑造数据的格式和结构,以确保其与目标系统和已有数据的兼容性。从逗号分隔的日志数据中解析字段以加载到关系数据库是此类数据转换的示例。

3.2 映射与转换

一些最基本的数据转换涉及数据的映射和转换。例如,包含表示错误代码的整数的列可以映射到相关的错误描述,使该列更易于理解,并且更便于在面向客户的应用程序中显示。

转换将数据从一个系统中使用的格式转换为适用于不同系统的格式。即使在解析之后,Web 数据也可能表示为分层 JSON 或 XML 文件的形式,但需要转换为行和列数据以存放在关系数据库中。

3.3 过滤、聚合和汇总

数据转换通常涉及减少数据并使其更易于管理。可以通过过滤掉不必要的字段、列和记录来合并数据。例如,通过将客户交易的时间序列转换为每小时或每天的销售计数来聚合或汇总数据。

商务智能工具可以进行这种过滤和聚合,但在报告工具访问数据之前,执行转换会更高效。

3.4 富集和填补

不同来源的数据可以合并来创建非规范化的丰富信息。客户的交易可以汇总到总计中,并添加到客户信息表中,以便更快地参考或供客户分析系统使用。长字段或自由格式字段可以被拆分为多个列,并且由于这些类型的转换,缺失值可能会被填补或损坏的数据被替换。

3.5 索引和排序

可以对数据进行转换,使其按逻辑顺序排列或适合数据存储方案。例如,在关系数据库管理系统中,创建索引可以提高性能或改进不同表之间关系的管理。

3.6 匿名化和加密

包含个人身份信息或其他可能危及隐私或安全的信息的数据应在传播前进行匿名处理。许多行业都要求对隐私数据进行加密,并且系统可以对从单个数据库单元到整个记录或字段的多个级别执行加密。

3.7 建模、类型转换、格式化和重命名

最后,一整套数据转换可以在不改变内容的情况下重塑数据。这包括为了兼容性转换数据类型、调整日期和时间,以及为清晰起见而重命名架构、表和列。

3.8 细化数据转换过程

在企业能够运行分析之前,甚至在转换数据之前,你必须将其复制到为分析而构建的

数据仓库中。如今，大多数组织都选择云数据仓库以充分利用 ELT。

Bigdata **Exercises**

〖Ex. 1〗 根据 Text A 回答以下问题。

1. What does data preprocessing involve?
2. What is the goal of data validation?
3. What does "garbage in, garbage out" mean?
4. What are the main goals of data quality assessment?
5. What can missing data do?
6. What is data cleansing?
7. What do data purging and data cleansing focus on respectively?
8. What is data integration?
9. What happens to databases when the volume of data is huge?
10. How does data deduplication work?

〖Ex. 2〗 根据 Text B 回答以下问题。

1. What is data transformation?
2. What may data transformation be?
3. What does properly formatted and validated data do?
4. What can performing transformations in an on-premises data warehouse do?
5. Why are data analysts without appropriate subject matter expertise less likely to notice typos or incorrect data?
6. What should the first phase of data transformations include?
7. What are initial transformations focused on?
8. What do some of the most basic data transformations involve?
9. What is data transformation often concerned with?
10. What can creating indexes do in relational database management systems?

〖Ex. 3〗 把下列单词或词组翻译成中文。

1. data cleaning   1. _____
2. data deduplication   2. _____
3. data cleansing   3. _____
4. data preprocessing   4. _____
5. data reduction   5. _____
6. data type   6. _____

7. garbage in, garbage out   7. _____

8. lossy reduction   8. _____

9. missing value   9. _____

10. standard deviation   10. _____

〖Ex. 4〗 把下列单词或词组翻译成英文。

1. 训练模型   1. _____

2. 数据整理   2. _____

3. 脚本语言   3. _____

4. n. 压缩   4. _____

5. v. 扰乱   5. _____

6. n. 分段   6. _____

7. n. 填补   7. _____

8. n. 不一致，不协调；前后矛盾   8. _____

9. n. 不合规；不规则；非正式   9. _____

10. n. 离群值；异常值   10. _____

〖Ex. 5〗 翻译句子。

1. This validation process ensures that the data conforms to acceptable formats.
2. Do these symbols have any particular significance?
3. Outlier data detection is an important part of data mining.
4. When there is no outlier in the sample, these methods can get good result.
5. Data cleansing is an important step both in data warehousing and data mining.
6. So, data cleaning is vital to improve data quality of information system.
7. The forecast database is established based on data cleaning and pretreatment.
8. Some data type changes may cause a change in the data.
9. You already know its name and data type.
10. Data compression has been a technique of indispensability in times of information.

Bigdata Reading Material

Big Data and Machine Learning

Big data and machine learning are the two hot topics in today's IT world. In this passage, we will study the relationship between the two.

1. What Is Machine Learning?

In simple words, we define machine learning as "Evolve through Learning".

Machine learning is the branch of computer science that allows machines to learn from past experiences[①] without explicitly doing the programming. It helps computers/machines in predicting the future without the intervention of humans. We can say that with the help of ML, software applications learn how to improve their accuracy to predict the outcomes. Hence, Machine learning enables machines to learn from data, find out useful hidden[②] patterns, and make decisions without human intervention.

2. Machine Learning in Big Data

Machine learning algorithms are useful for data collection, data analysis, and data integration. ML algorithms are a must for the larger organizations which are generating tons[③] of data.

We can apply ML algorithms to every element of big data operation, including:

- Data labeling[④] and segmentation
- Data analytics
- Scenario simulation[⑤]

All these stages are integrated for generating insights, patterns, which are then categorized and packaged into an easily understandable format. The fusion of big data with machine learning is a never-ending[⑥] loop[⑦].

3. Relationship Between Big Data and Machine Learning

It is always better to have several varieties of data to get them filtered[⑧] for generating accurate results. But managing these wide varieties of data is very difficult. So it becomes a challenge to manage and analyze big data. Also, information is useless[⑨] until it is well interpreted.

---

① experience [ɪkˈspɪərɪəns] n. 经验，经历

② hidden [ˈhɪdn] adj. 隐藏的

③ ton [tʌn] n. 大量，许多

④ data labeling：数据标记

⑤ scenario simulation：情景模拟

⑥ never-ending [ˈnevəˌendɪŋ] adj. 无止境的；无限的

⑦ loop [luːp] n. 循环

⑧ filter [ˈfɪltə] n. 过滤器；筛选(过滤)程序 v. 过滤；筛选

⑨ useless [ˈjuːsləs] adj. 无用的，无效的

Thus, to use information, there is a need for talent①, algorithms, and computing infrastructure.

Machine learning enables machines to use the data provided by big data and respond accurately, thus leading to improved service quality, business operations, customer relationships, and more.

Machine learning algorithms take data from big data and learn more. Big data analytics provide varieties of data to the machines to show and give better results. Thus, Businesses can fulfill their dreams and get the advantages of big data by using machine learning algorithms but with the help of skilled data scientists to run that data into knowledge.

4. Difference Between Big Data and Machine Learning

Big data is related to data storage, ingestion and extraction tools such as Apache Hadoop, Spark, etc. whereas machine learning is a subset of AI that enables machines to predict the future without human intervention②.

Big data is the analysis of vast amounts of data by discovering useful hidden patterns or extracting information from it. Thus, big data is huge information analytics where we perform analysis on huge information. While machine learning teaches computers to take input data and give desired outputs based on the machine learning models.

Big data analytics is all about collecting and transforming raw data③ into extracted information, and this data information is then used by the machine learning algorithms to predict better results.

Machine learning is a part of data science while big data is related to high-performance④ computing.

Machine learning processes data and generates output without human intervention whereas big data analysis involves human interaction.

We can set up both machine learning and big data to automatically look for particular types of data, parameters and relationship between them. But big data can not see the relationship between existing pieces of data and parameters with the same depth as machine learning.

5. Big Data and Machine Learning Use Cases

The fusion of machine learning and big data is the reason behind the growth of many industries. Here we have listed some use cases of machine learning and big data.

5.1 Market research and target audience segmentation

In order to gain profits, knowing the audience is one of the most critical elements for a

① talent ['tælənt] n. 人才
② intervention [ˌɪntə'venʃn] n. 干预，介入，干涉
③ raw data：原始数据
④ high-performance [haɪ-pə'fɔ:məns] adj. 高性能的

profitable[1] business. Machine learning algorithms study the market and help business organizations to understand their target audience.

By using supervised and unsupervised ML algorithms, organizations can find out the portrait[2] of their target audience, patterns of their behavior, and their preferences. This technique is used in e-commerce[3], media & entertainment, advertising, and many other domains.

5.2 User modeling

It is an elaboration[4] on target audience segmentation. User modeling dwells[5] inside the user behavior and creates a detailed portrait of a specific segment. By using machine learning algorithms for big data analytics, we can predict users' behavior and make intelligent business decisions.

Facebook is an example of such a user modeling system. It creates a detailed portrait of the user for suggesting friends, pages, communities, ads, etc.

5.3 Recommendation engine

Recommendation engine[6] is the best use case of big data with machine learning. This system provides the best suggestions[7] for the types of products to be brought together, contents the user might be interested to read or see.

Based on the combination[8] of context and user behavior prediction, this system can shape user experience according to the user's expressed preferences and behavior on the site.

Recommendation engines apply content-based data filtering for extracting insights. Thus, the system learns from the user's preferences and tendencies.

Amazon and Netflix popularly use recommendation engines.

5.4 Predictive analysis

Big data with machine learning plays a vital role in shaping the bright future of retail[9] industries. For retail, knowing customers' needs is one of the most important elements. Thus they use market basket analysis.

Big data allows retailers to calculate the probabilities of different outcomes and decisions. Predictive analytics helps them by providing suggestions for extra products on e-commerce

---

① profitable ['prɒfɪtəbl] adj. 有利可图的，有益的；可赚钱的

② portrait ['pɔːtreɪt] n. 肖像；模型，标本

③ e-commerce [iː'kɒmɜːs] n. 电子商务

④ elaboration [ɪˌlæbə'reɪʃn] n. 精心制作；详尽阐述；苦心经营

⑤ dwell [dwel] v. 居住，栖身

⑥ recommendation engine：推荐引擎

⑦ suggestion [sə'dʒestʃn] n. 建议

⑧ combination [ˌkɒmbɪ'neɪʃn] n. 组合，结合

⑨ retail [rɪ'teɪl] n. & vt. 零售；零卖

platforms.

eBay's system is an example of predictive analysis that reminds us of abandoned purchases, incoming auctions①, or hot deal②s.

5.5　Chatbots

Chatbots also are known as the conversational③ user interface is another most important use case of big data with machine learning. By using machine learning algorithms, a chatbot can easily adapt to a particular customer's preferences④ after interactions.

Amazon's Alexa and Apple's Siri are the most well known AI assistants⑤.

6.　Summary

In short, we can say that big data and machine learning are different from each other but these two hot trending technologies are used in combination for a successful business.

The input to the machine learning algorithms is the information extracted by big data analysis. This input is then learned by the machine learning models to predict desired outputs.

---

① auction ['ɔːkʃn]　n. 拍卖(会)　v. 拍卖
② hot deal：热门交易，热门买卖
③ conversational [ˌkɒnvə'seɪʃənl]　adj. 会话的，谈话的
④ preference ['prefrəns]　n. 偏爱；优先权
⑤ assistant [ə'sɪstənt]　n. 助手，助理

Unit 5

ELT

In a data-driven world, an efficient process for moving and transforming data for analysis is critical to business growth and innovation. Loading a data warehouse can be an extremely time-consuming process. The process of extracting, loading, and transforming data streamlines the tasks of modern data warehousing and managing big data so that businesses can focus on mining their data for actionable insights.

ELT is the process of extracting data from one or multiple sources and loading it into a target data warehouse. Instead of transforming the data before it's written, ELT takes advantage of the target system to do the data transformation. This approach requires fewer remote sources than other techniques because it needs only raw and unprepared data.

ELT is an alternative to the traditional extract, transform, load process. It pushes the transformation component of the process to the target database for better performance. This capability is very useful for processing the massive data sets needed for business intelligence (BI) and big data analytics.

Because it takes advantage of the processing capability already built into a data storage infrastructure, ELT reduces the time data spends in transit and boosts efficiency.

Though the ELT process has been in practice for some time, it is gaining popularity with the more widespread use of Hadoop and cloud-native data lakes.

1. How ELT Works?

It is becoming increasingly common for data to be extracted from its source locations, then loaded into a target data warehouse to be transformed into actionable business intelligence. This process consists of three steps:

(1) Extract—This step works similarly in both ETL and ELT data management approaches. Raw streams of data from virtual infrastructure, software, and applications are ingested either in their entirety or according to predefined rules.

(2) Load—Here is where ELT branches off from its ETL cousin. Rather than deliver this mass of raw data and load it to an interim processing server for transformation, ELT delivers it directly to the target storage location. This shortens the cycle between extraction and delivery.

(3) Transform—The database or data warehouse sorts and normalizes the data and keep part or all of it on hand and accessible for customized reporting. The overhead for storing this much data is higher, but it offers more opportunities to mine it for relevant business intelligence in near real-time.

2. Benefits of ELT

The explosion in the types and volume of data that businesses must process can put a strain on traditional data warehouses. ELT offers a number of advantages, including:

*   Simplifying management—ELT separates the loading and transformation tasks, minimizing the interdependencies between these processes, lowering risk and streamlining project management.
*   Future-proofed data sets—ELT implementations can be used directly for data warehousing systems, but ELT is often used in the data lake approach in which data is collected from a range of sources. This, combined with the separation of the transformation process, makes it easier to make future changes to the warehouse structure.
*   Leveraging the latest technologies—ELT solutions harness the power of new technologies in order to push improvements, security and compliance across the enterprise. ELT also leverages the native capabilities of modern cloud data warehouses and big data processing frameworks.
*   Lowering costs—Like most cloud services, cloud-based ELT can result in lower total cost of ownership because an upfront investment in hardware is often unnecessary.
*   Flexibility—The ELT process is adaptable and flexible, so it's suitable for a variety of businesses, applications, and goals.
*   Scalability—The scalability of a cloud infrastructure and hosted services like platform as a service (PaaS) and software as a service (SaaS) give organizations the ability to expand resources on the fly. They add the compute time and storage space necessary for massive data transformation tasks.

Although it is still evolving, ELT offers the promise of unlimited access to data, less development time, and significant cost savings. In these ways, the cloud is redefining data integration.

3. Typical Capabilities of ETL Tools

In the past, organizations wrote their own ETL code. There are now many open source and commercial ETL tools and cloud services to choose from. Typical capabilities of these products include the following:

*   Comprehensive automation and ease of use: Leading ETL tools automate the entire data

flow from data sources to the target data warehouse. Many tools recommend rules for extracting, transforming and loading the data.

• A visual, drag-and-drop interface: This functionality can be used for specifying rules and data flows.

• Support for complex data management: This includes assistance with complex calculations, data integrations and string manipulations.

• Security and compliance: The best ETL tools encrypt data both in motion and at rest and are certified compliant with industry or government regulations, like HIPAA and GDPR.

In addition, many ETL tools have evolved to include ELT capability and to support integration of real-time and streaming data for artificial intelligence (AI) applications (see Figure 5-1).

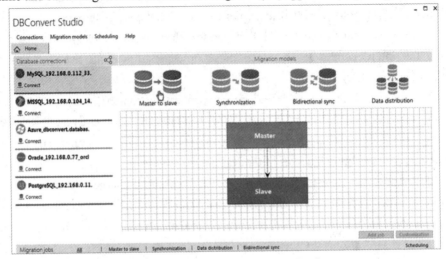

Figure 5-1   ETL Tool: DB Convert Studio

4. ETL vs. ELT: Understanding the Differences

The primary differences between ETL and ELT are how much data is retained in data warehouses and where data is transformed. With ETL, the transformation of data is done before it is loaded into a data warehouse. This enables analysts and business users to get the data they need faster, without building complex transformations or persistent tables in their business intelligence tools. Using the ELT approach, data is loaded into the warehouse or data lake as is, with no transformation before loading. This makes jobs easier to configure because it only requires an origin and a destination.

The ETL and ELT approaches to data integration differ in several key ways:

• Load time—It takes significantly longer to get data from source systems to the target system with ETL.

• Transformation time—ELT performs data transformation on-demand using the target system's computing power, thus reducing wait times for transformation.

• Complexity—ETL tools typically have an easy-to-use GUI that simplifies the process. ELT requires in-depth knowledge of BI tools, masses of raw data and a database that can transform it effectively.

• Data warehouse support—ETL is a better fit for legacy on-premise data warehouses and structured data. ELT is designed for the scalability of the cloud.

• Maintenance—ETL requires significant maintenance for updating data in the data warehouse. With ELT, data is always available in near real-time.

Both ETL and ELT processes have their place in today's competitive landscape, and understanding a business' unique needs and strategies is key to determining which process will deliver the best results.

5. ELT and Data Lakes: the Future of Data Integration?

Modern cloud-based infrastructure technologies offer large amounts of data storage and scalable computing power at lower costs, making it possible to keep petabytes of data in large and expandable data lakes, and process it quickly on-demand. The proliferation of data lakes has made it possible for more organizations to move from ETL to ELT.

Data lakes offer major benefits for organizations migrating big data and big data processes from on-premises to the cloud. They enable data to be kept in a more flexible format for future use, along with identifiers and metadata tags for faster retrieval. They also offer fast load time.

ELT appears to be the future of data integration. It offers many advantages over ETL, which is an older, slower process. Data volume has grown exponentially for organizations, and ETL tools cannot efficiently handle the integration of all this data into a repository for analysis. ELT delivers better agility and less maintenance, making it a cost-effective way for businesses of all sizes to take advantage of cloud-based data storage such as data lakes.

Data plays a critical role in every business operation. In order to be valuable, it must be moved and prepared for use. ELT is an essential piece of the data integration process, providing a different approach to data movement than the traditional ETL process.

New Words

innovation	[ˌɪnəˈveɪʃn]	n. 改革，创新；新观念；新发明
extremely	[ɪkˈstriːmli]	adv. 极端地；非常，很
target	[ˈtɑːgɪt]	n. 目标
unprepared	[ˌʌnprɪˈpeəd]	adj. 无准备的；尚未准备好的
alternative	[ɔːlˈtɜːnətɪv]	n. 可供选择的事物 adj. 替代的，备选的
component	[kəmˈpəʊnənt]	n. 组成部分，成分 adj. 组成的，构成的
transit	[ˈtrænzɪt]	n. 传输；运输
boost	[buːst]	v. 提高，增强，推动

widespread	['waɪdspred]	adj. 分布广的，普遍的
gain	[geɪn]	v. 获得；受益
		n. 增加；好处
entirety	[ɪn'taɪərəti]	n. 全部，整体
rule	[ruːl]	n. 规则，行事准则
		v. 控制，统治，支配
deliver	[dɪ'lɪvə]	v. 交付，递送
interim	['ɪntərɪm]	adj. 暂时的，临时的
shorten	['ʃɔːtn]	v. (使)变短，缩短
cycle	['saɪkl]	n. 周期；循环
explosion	[ɪk'spləʊʒn]	n. 激增；爆发
separate	['sepəreɪt]	v. (使)分开，分离；分割，划分
	['sepərɪt]	adj. 独立的，分开的；不同的，不相关的
minimize	['mɪnɪmaɪz]	v. 最小化，把……减至最低数量
interdependency	[ɪntə'dɪpendənsi]	n. 相关性，相依性
leverage	['liːvərɪdʒ]	v. 利用
harness	['haːnɪs]	v. 利用；控制
framework	['freɪmwɜːk]	n. 构架；框架；(体系的)结构
adaptable	[ə'dæptəbl]	adj. 可适应的；有适应能力的
redefine	[ˌriːdɪ'faɪn]	v. 重新定义，再定义
comprehensive	[ˌkɒmprɪ'hensɪv]	adj. 全面的；综合性的
drag-and-drop	[dræg ænd drɔp]	adj. (鼠标等)拖放的
functionality	[ˌfʌŋkʃə'næləti]	n. 功能性
assistance	[ə'sɪstəns]	n. 帮助；支持；辅助
regulation	[ˌregjʊ'leɪʃn]	n. 规章，规则
		adj. 规定的
real-time	['riːl 'taɪm]	adj. 实时的，即时处理的
competitive	[kəm'petətɪv]	adj. 竞争的
expandable	[ɪk'spændəbl]	adj. 可扩展的，可扩大的
exponentially	[ˌekspə'nenʃəli]	adv. 以指数方式
essential	[ɪ'senʃl]	adj. 基本的，根本的；必不可少的

Phrases

| data transformation | 数据转换 |
| remote source | 远程资源 |

target database	目标数据库，目的数据库
be built into	内置
data storage infrastructure	数据存储基础设施
source location	源位置
virtual infrastructure	虚拟基础设施
on hand	在手头，现有
customized reporting	自定义报告
project management	项目管理
a range of	一系列
total cost of ownership	总拥有成本
upfront investment	前期投资
host service	托管服务
storage space	存储空间
string manipulation	字符串处理
load time	装入时间，存入时间
computing power	计算能力
metadata tag	元数据标签
cloud-based data storage	基于云的数据存储

Abbreviations

PaaS (Platform as a Service)	平台即服务
HIPAA (Health Insurance Portability and Accountability Act)	健康保险携带和责任法案
GDPR (General Data Protection Regulation)	通用数据保护条例
GUI (Graphical User Interface)	图形用户界面

Analysis of Difficult Sentences

[1] Because it takes advantage of the processing capability already built into a data storage infrastructure, ELT reduces the time data spends in transit and boosts efficiency.

　　本句中，Because it takes advantage of the processing capability already built into a data storage infrastructure 是原因状语从句，修饰谓语 reduces 和 boosts。在该从句中，already built into a data storage infrastructure 过去分词短语，作定语，修饰和限定 the processing capability。data spends in transit 是定语从句，修饰和限定 the time。take advantage of 的意思是"利用"。

[2] The scalability of a cloud infrastructure and hosted services like platform as a service

(PaaS) and software as a service (SaaS) give organizations the ability to expand resources on the fly.

本句中，The scalability of a cloud infrastructure and hosted services 是名词短语，作主语。like integration platform as a service (PaaS) and software as a service (SaaS)对 hosted services 举例说明，谓语是 give，to expand resources on the fly 是动词不定式短语，作定语，修饰和限定 the ability。on the fly 的意思是"及时地，动态地"。

[3]　This enables analysts and business users to get the data they need faster, without building complex transformations or persistent tables in their business intelligence tools.

本句中，they need faster 是定语从句，修饰和限定 the data。

[4]　Modern cloud-based infrastructure technologies offer large amounts of data storage and scalable computing power at lower costs, making it possible to keep petabytes of data in large and expandable data lakes, and process it quickly on-demand.

本句中，making it possible to keep petabytes of data in large and expandable data lakes, and process it quickly on-demand 是现在分词短语，作结果状语。在该短语中，to keep petabytes of data in large and expandable data lakes, and process it quickly on-demand 是动词不定式短语，作 make 的真正的宾语，make 后面 it 是形式宾语。

 参考译文

提取、加载和转换

在数据驱动的世界中，移动和转换数据以进行分析的高效流程对于业务增长和创新至关重要。加载数据仓库是非常耗时的过程。提取、加载和转换数据的过程简化了现代数据仓库和管理大数据的任务，以便企业可以专注于挖掘数据来获得可行动的见解。

ELT 是从一个或多个源提取数据并将其加载到目标数据仓库的过程。ELT 不是在写入数据之前转换数据，而是利用目标系统进行数据转换。这种方法比其他技术耗费更少的远程资源，因为它只需要原始和未准备好的数据。

ELT 是传统提取、转换、加载过程的替代方案。它将流程的转换组件推送到目标数据库以获得更好的性能。此功能对于处理商业智能 (BI) 和大数据分析所需的海量数据集非常有用。

因为 ELT 利用了数据存储基础设施中内置的处理能力，它减少了数据在传输过程中花费的时间并提高了效率。

尽管 ELT 流程已经在实践中使用了一段时间，但随着 Hadoop 和云原生数据湖的广泛使用，它越来越受欢迎。

1. ELT 的工作原理

从源位置提取数据，然后将其加载到目标数据仓库中以转换为可操作的商业智能，这

种情况越来越普遍。这个过程包括以下三个步骤：

(1) 提取——这一步在 ETL 和 ELT 数据管理方法中的工作方式类似。来自虚拟基础设施、软件和应用程序的原始数据流被整体或根据预定义的规则获取。

(2) 加载——这里是 ELT 从 ETL 分支出来的地方。ELT 不是交付大量原始数据并将其加载到临时处理服务器进行转换，而是直接将其交付到目标存储位置。这缩短了提取和交付之间的周期。

(3) 转换——数据库或数据仓库对数据进行分类和规范化，并将部分或全部数据保留在手边，并可用于自定义报告。存储这么多数据的开销更高，但它为近乎实时的方式挖掘相关商业智能提供了更多机会。

2. ELT 的好处

企业必须处理的数据类型和数量的爆炸式增长会给传统数据仓库带来压力。ELT 提供了许多优势，包括：

• 简化管理——ELT 将加载和转换任务分开，最大限度地减少这些过程之间的相互依赖性，降低了风险，并简化了项目管理。

• 面向未来的数据集——ELT 可直接用于数据仓库系统，但也经常用于多来源收集数据的数据湖。与转换过程的分离相结合，使改变仓库结构更加容易。

• 利用最新技术——ELT 解决方案利用新技术的力量来推动整个企业的改进、安全性和合规性。ELT 还利用了现代云数据仓库和大数据处理框架的固有功能。

• 降低成本——与大多数云服务一样，因为通常不需要对硬件进行前期投资，基于云的 ELT 可以降低总拥有成本。

• 灵活性——ELT 流程具有适应性和灵活性，因此适用于各种业务、应用程序和目标。

• 可扩展性——云基础架构和托管服务(如平台即服务(iPaaS)和软件即服务(SaaS))的可扩展性使组织能够动态扩展资源。它们增加了海量数据转换任务所需的计算时间和存储空间。

尽管 ELT 仍在不断发展，但它提供了对数据的无限访问、缩短了开发时间并节省了大量成本。通过这些方式，云正在重新定义数据集成。

3. ETL 工具的典型能力

过去，组织编写自己的 ETL 代码。现在有许多开源和商业 ETL 工具和云服务可供选择。这些产品的典型功能包括：

• 全面的自动化和易用性：领先的 ETL 工具能够自动化整个数据流，从数据源到目标数据仓库。许多工具推荐于提取、转换和加载数据的规则。

• 可视化的拖放界面：此功能可用于指定规则和数据流。

• 支持复杂的数据管理：这包括协助进行复杂的计算、数据集成和字符串操作。

• 安全性和合规性：最好的 ETL 工具可以对动态和静态数据进行加密，并且经过认证符合行业或政府法规，例如 HIPAA 和 GDPR。

此外，许多 ETL 工具已经发展到包括 ELT 功能并支持为人工智能(AI)应用集成实时和流式数据(见图 5-1)。

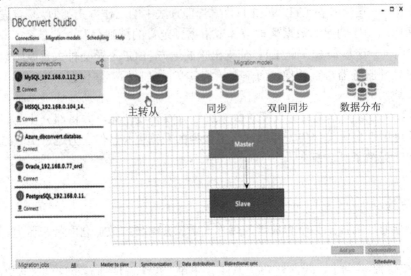

图 5-1　ELT 工具：DB 转换工作室

4．ETL 和 ELT：理解差异

ETL 和 ELT 之间的主要区别在于数据仓库中保留了多少数据以及数据在哪里转换。使用 ETL，数据的转换是在将数据加载到数据仓库之前完成的。这使分析师和业务用户能够更快地获得他们需要的数据，而无需在商业智能工具中构建复杂的转换或持久的表。使用 ELT 方法，数据按原样加载到仓库或数据湖中，加载前不进行任何转换。因为它只需要一个源点和一个目的地，这样做更容易配置。

数据集成的 ETL 和 ELT 方法在几个关键方面有所不同：

- 加载时间——使用 ETL 将数据从源系统传输到目标系统所需的时间明显更长。
- 转换时间——ELT 使用目标系统的计算能力按需执行数据转换，从而减少了转换等待时间。
- 复杂性——ETL 工具通常具有易于使用的 GUI，可简化流程。ELT 需要深入了解商务智能工具、大量原始数据以及可以有效转换数据的数据库。
- 数据仓库支持——ETL 更适合传统的内部部署数据仓库和结构化数据。ELT 专为云的可扩展性而设计。
- 维护——ETL 需要大量维护以更新数据仓库中的数据。使用 ELT，数据几乎总是实时可用。

ETL 和 ELT 流程在当今的竞争格局中都占有一席之地，了解企业的独特需求和战略是确定哪个流程将提供最佳结果的关键。

5．ELT 和数据湖：数据集成的未来？

现代的基于云的基础设施技术以较低的成本提供了大量数据存储和可扩展的计算能力，从而可以将 PB 级数据保存在可扩展的大型数据湖中，并按需快速处理。数据湖的激增使更多组织从 ETL 转向 ELT 成为可能。

数据湖主要为将大数据和大数据流程从本地迁移到云的组织提供了好处。它们使数据能够以更灵活的格式保存以备将来使用，同时提供了标识符和元数据标签来加快检索速度，也节省了加载时间。

ELT 似乎是数据集成的未来。与 ETL 相比，它有许多优势，ETL 是一个较旧、较慢的过程。组织的数据量呈指数级增长，而 ETL 工具无法有效地将所有数据集成到存储库中进行分析。ELT 敏捷性更高，维护更少，这使其成为各种规模的企业利用基于云的数据存储(例如数据湖)的经济高效的方式。

数据在每项业务运营中都发挥着至关重要的作用。为了使其有价值，必须将其移动并备用。ELT 是数据集成过程的重要组成部分，它提供了一种不同于传统 ETL 过程的数据移动方法。

Bigdata Text B

Big Data Analytics

Each day, your customers generate an abundance of data. Every time they open your email, use your mobile app, tag you on social media, walk into your store, make an online purchase, talk to a customer service representative, or ask a virtual assistant about you, those technologies collect and process that data for your organization. And that's just your customers. Each day, employees, supply chains, marketing efforts, finance teams and more generate an abundance of data, too. Big data is an extremely large volume of data and datasets that come in diverse forms and from multiple sources. Many organizations have recognized the advantages of collecting as much data as possible. But it's not enough just to collect and store big data—you also have to put it to use. Thanks to rapidly growing technology, organizations can use big data analytics to transform terabytes of data into actionable insights.

1. What Is Big Data Analytics?

Big data analytics describes the process of uncovering trends, patterns, and correlations in large amounts of raw data to help make data-informed decisions. These processes use familiar statistical analysis techniques—like clustering and regression—and apply them to more extensive datasets with the help of newer tools. Big data has been a buzz word since the early 2000s, when software and hardware capabilities made it possible for organizations to handle large amounts of unstructured data. Since then, new technologies—from Amazon to smartphones —have contributed even more to the substantial amounts of data available to organizations. With the explosion of data, early innovation projects like Hadoop, Spark, and NoSQL databases were created for the storage and processing of big data. This field continues to evolve as data engineers look for ways to integrate the vast amounts of complex information created by sensors, networks, transactions, smart devices, web usage and more. Even now, big data analytics methods are being used with emerging technologies, like machine learning, to discover and scale more complex insights.

2. How Big Data Analytics Works?

Big data analytics refers to collecting, processing, cleaning and analyzing large datasets to help organizations operationalize their big data.

2.1 Collect data

Data collection looks different for every organization. With today's technology, organizations can gather both structured and unstructured data from a variety of sources—from cloud storage to mobile applications to in-store IoT sensors and beyond. Some data will be stored in data warehouses where business intelligence tools and solutions can access it easily. Raw or unstructured data that is too diverse or complex for a warehouse may be assigned metadata and stored in a data lake.

2.2 Process data

Once data is collected and stored, it must be organized properly to get accurate results on analytical queries, especially when it's large and unstructured. Available data is growing exponentially, making data processing a challenge for organizations. One processing option is batch processing, which looks at large data blocks over time. Batch processing is useful when there is a longer turnaround time between collecting and analyzing data. **Stream** processing looks at small batches of data at once, shortening the delay time between collection and analysis for quicker decision-making. Stream processing is more complex and often more expensive.

2.3 Clean data

Data big or small requires scrubbing to improve data quality and get better results. All data must be formatted correctly, and any duplicative or irrelevant data must be eliminated. Dirty data can obscure and mislead, creating flawed insights.

2.4 Analyze data

Getting big data into a usable state takes time. Once it's ready, advanced analytics processes can turn big data into big insights. Some of these big data analysis methods include:

• Data mining sorts through large datasets to identify patterns and relationships by identifying anomalies and creating data clusters.

• Predictive analytics uses an organization's historical data to make predictions about the future, identifying upcoming risks and opportunities.

• Deep learning finds patterns in the most complex and abstract data.

3. Big Data Analytics Tools and Technology

Big data analytics cannot be narrowed down to a single tool or technology. Instead, several types of tools work together to help you collect, process, cleanse and analyze big data. Some of the major players in big data ecosystems are listed below.

• Hadoop is an open-source framework that efficiently stores and processes big datasets on clusters of commodity hardware. This framework is free and can handle large amounts of structured and unstructured data, making it a valuable mainstay for any big data operation.

• NoSQL databases are non-relational data management systems that do not require a fixed scheme, making them a great option for big, raw, unstructured data. NoSQL stands for "not only SQL," and these databases can handle a variety of data models.

• MapReduce is an essential component to the Hadoop framework serving two functions. The first is mapping, which filters data to various nodes within the cluster. The second is reducing, which organizes and reduces the results from each node to answer a query.

• YARN stands for "Yet Another Resource Negotiator." It is another component of second-generation Hadoop. The cluster management technology helps with job scheduling and resource management in the cluster.

• Spark is an open-source cluster computing framework that uses implicit data parallelism and fault tolerance to provide an interface for programming entire clusters. Spark can handle both batch and stream processing for fast computation (see Figure 5-2).

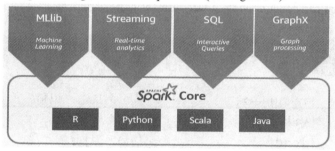

Figure 5-2   Spark Core

• Tableau is an end-to-end data analytics platform that allows you to prep, analyze, collaborate, and share your big data insights. Tableau excels in self-service visual analysis, allowing people to ask new questions of governed big data and easily share those insights across the organization.

4. The Big Benefits of Big Data Analytics

The ability to analyze more data at a faster rate can provide big benefits to an organization, allowing it to more efficiently use data to answer important questions. Big data analytics is important because it lets organizations use colossal amounts of data in multiple formats from multiple sources to identify opportunities and risks, helping organizations move quickly and improve their bottom lines. Some benefits of big data analytics include:

• Cost savings. Helping organizations identify ways to do business more efficiently.

• Product development. Providing a better understanding of customer needs.

• Market insights. Tracking purchase behavior and market trends.

5. The Big Challenges of Big Data

Big data brings big benefits, but it also brings big challenges such new privacy and security concerns, accessibility for business users and choosing the right solutions for your business needs. To capitalize on incoming data, organizations will have to address the following:

• Making big data accessible. Collecting and processing data becomes more difficult as the amount of data grows. Organizations must make data easy and convenient for data owners of all skill levels to use.

• Maintaining quality data. With so much data to maintain, organizations are spending more time than ever before scrubbing for duplicates, errors, absences, conflicts, and inconsistencies.

• Keeping data secure. As the amount of data grows, so do privacy and security concerns. Organizations will need to strive for compliance and put tight data processes in place before they take advantage of big data.

• Finding the right tools and platforms. New technologies for processing and analyzing big data are developed all the time. Organizations must find the right technology to work within their established ecosystems and address their particular needs. Often, the right solution is also a flexible solution that can accommodate future infrastructure changes.

New Words

recognize	['rekəgnaɪz]	v. 认识，认出；识别；承认
uncover	[ʌn'kʌvə]	v. 揭示
trend	[trend]	n. 趋势，倾向
regression	[rɪ'greʃn]	n. 回归
smartphone	['smɑːtfəʊn]	n. 智能手机
continue	[kən'tɪnjuː]	v. 连续，持续，继续
operationalize	[ɒpə'reɪʃənlaɪz]	v. 使用于操作，实施
solution	[sə'luːʃn]	n. 解决办法，解决方案
assign	[ə'saɪn]	v. 分配，指派(任务)
turnaround	['tɜːnəraʊnd]	n. 周转
scrub	[skrʌb]	v. 用力擦洗，刷洗
obscure	[əb'skjʊə]	v. 使……模糊不清；隐藏，掩盖 adj. 不清楚的；隐蔽的
mislead	[ˌmɪs'liːd]	v. 误导；将……引入歧途
flawed	[flɔːd]	adj. 有缺点的；有缺陷的；错误的
usable	['juːzəbl]	adj. 可用的，合用的，便于使用的
state	[steɪt]	n. 状态
imitate	['ɪmɪteɪt]	v. 模仿，效仿

102

ecosystem	[ˈiːkəʊsɪstəm]	n. 生态系统
mainstay	[ˈmeɪnsteɪ]	n. 支柱，骨干
second-generation	[ˈsekənd ˌdʒenəˈreiʃn]	n. 第二代
parallelism	[ˈpærəlelɪzəm]	n. 平行
end-to-end	[endtuːend]	adj. 端到端的，端对端的
prep	[prep]	v. 预备，准备 n. 准备 adj. 预备的
collaborate	[kəˈlæbəreɪt]	v. 合作，协作
self-service	[self ˈsɜːvɪs]	adj. 自我服务的；(商店)自选的；(饭店)自助的
rate	[reɪt]	n. 速度，进度；比率
colossal	[kəˈlɒsl]	adj. 巨大的
capitalize	[ˈkæpɪtəlaɪz]	v. 利用 v. 估计……的价值
accessible	[əkˈsesəbl]	adj. 可访问的
absence	[ˈæbsəns]	n. 缺失，缺乏
conflict	[ˈkɒnflɪkt]	n.&v. 冲突
accommodate	[əˈkɒmədeɪt]	v. 顾及，考虑到；适应

Phrases

an abundance of ...	大量的……
online purchase	在线购物
virtual assistant	虚拟助理
supply chain	供应链
emerging technology	新兴技术；新兴科技
cloud storage	云存储
mobile application	移动应用
business intelligence tool	商务智能工具
batch processing	批处理
data block	数据块
delay time	延迟时间
be narrowed down to	缩小到
non-relational data management system	非关系性数据管理系统
stand for	代表
stream processing	流处理
bottom line	底线

| market trend | 市场趋势 |
| strive for | 争取 |

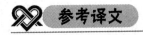

大 数 据 分 析

每天，你的客户都会生成大量数据。每次他们打开电子邮件、使用移动应用程序、在社交媒体上标记、走进商店、线上购物、与客户服务代表交谈或向虚拟助理询问情况时，这些技术都会为你的组织收集并处理这些数据。这些只是你的客户。每天，员工、供应链、营销工作、财务团队等也会生成大量数据。大数据是大量数据和数据集，它形式多样，来源广泛。许多组织已经认识到尽可能多地收集数据的优势。但仅仅收集和存储大数据是不够的——还必须将其投入使用。得益于快速发展的技术，组织可以使用大数据分析将 TB 级数据转化为可行动的见解。

1. 什么是大数据分析？

大数据分析描述了在大量原始数据中发现趋势、模式和相关性以帮助做出基于数据的决策过程。这些过程使用熟悉的统计分析技术——如聚类和回归——并在更新的工具的帮助下将其应用于更广泛的数据集。自 21 世纪初以来，大数据一直是一个流行词，当时软件和硬件功能使组织可以处理大量非结构化数据。从那时起，新技术——从亚马逊到智能手机——为组织提供大量数据做出了更大的贡献。随着数据的爆炸式增长，早期创新项目 Hadoop、Spark 和 NoSQL 数据库等被创建起来用于大数据的存储和处理。随着数据工程师寻找方法来集成由传感器、网络、交易、智能设备、网络使用等产生的大量复杂信息，该领域不断发展。即使是现在，大数据分析方法也正在与机器学习等新兴技术结合使用，来发现和扩展更复杂的见解。

2. 大数据分析的工作原理

大数据分析是指收集、处理、清理和分析大型数据集，以帮助组织运营其大数据。

2.1 收集数据

每个组织的数据收集看起来都不一样。借助当今的技术，组织可以从各种来源收集结构化和非结构化数据——从云存储到移动应用程序，再到店内物联网传感器等等。一些数据将存储在数据仓库中，商业智能工具和解决方案可以轻松访问这些数据。对于仓库而言，过于多样化或复杂的原始或非结构化数据可能会被分配元数据并存储在数据湖中。

2.2 处理数据

一旦收集和存储了数据，就必须对其进行适当的组织以便在分析查询中获得准确的结果，尤其是在数据庞大且非结构化的情况下。可用数据正在呈指数级增长，这让数据处理成为组织面临的挑战。一种处理方法是批处理，随着时间的推移查看大数据块。当收集和分析数据之间的周转时间较长时，批处理很有用。流处理可以一次查看小批量数据，这缩短了收集和分析之间的延迟时间，可以更快地做出决策。流处理更复杂，通常也更昂贵。

2.3　清理数据

大数据或小数据都需要清理来提高数据质量并获得更好的结果。所有数据都必须正确格式化，并且必须消除重复或不相关的数据。"脏"数据可能会模糊和误导，从而产生有缺陷的见解。

2.4　分析数据

让大数据进入可用状态需要时间。一旦准备就绪，高级分析流程可以将大数据转化为非常有用的见解。大数据分析方法包括：

- 数据挖掘对大型数据集进行排序，通过识别异常和创建数据集群来识别模式和关系。
- 预测分析会使用组织的历史数据来预测未来，识别即将到来的风险和机遇。
- 深度学习可以在最复杂和抽象的数据中找到模式。

3.　大数据分析工具和技术

不能把大数据分析缩小为单一工具或技术。相反，应使用多种类型的工具协同工作来帮助收集、处理、清理和分析大数据。下面列出了大数据生态系统中的一些主要参与者。

- Hadoop 是一种开源框架，可在商用硬件集群上高效存储和处理大型数据集。该框架是免费的，可以处理大量结构化和非结构化数据，使其成为大数据操作的宝贵支柱。
- NoSQL 数据库是非关系型数据管理系统，不需要固定方案，是处理大型原始非结构化数据的绝佳选择。NoSQL 代表"不仅是 SQL"，这些数据库可以处理各种数据模型。
- MapReduce 是 Hadoop 框架的重要组件，可提供两个功能。第一个是映射，它将数据过滤到集群内的各个节点。第二个是减少，它组织和减少每个节点的结果来回答查询。
- YARN 代表"Yet Another Resource Negotiator(又一个资源协调者)"。它是第二代 Hadoop 的另一个组件。集群管理技术有助于集群中的作业调度和资源管理。
- Spark 是一个开源集群计算框架，它使用隐式数据并行性和容错性为整个集群编程提供接口。批处理和流处理均可以通过 Spark 进行处理以实现快速计算(见图 5-2)。

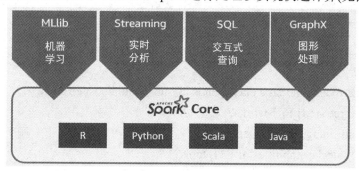

图 5-2　Spack Core

- Tableau 是一个端到端的数据分析平台，可让你准备、分析、协作和分享大数据见解。Tableau 擅长自助式可视化分析，允许人们对被管理的大数据提出新的问题，并很容易地在整个组织中分享这些见解。

4．大数据分析的巨大益处

大数据分析以更快的速度分析更多数据的能力可以为组织带来巨大好处，使其能够更有效地使用数据来回答重要问题。大数据分析很重要，因为它让组织可以使用来自多个来源的多种格式的大量数据来识别机会和风险，帮助组织快速行动并提高他们的底线。大数据分析的益处包括：

- 节约成本。帮助组织确定更有效地开展业务的方法。
- 产品开发。提供对客户需求的更好理解。
- 市场洞察。跟踪购买行为和市场趋势。

5．大数据的巨大挑战

大数据带来了巨大的好处，但也带来了巨大的挑战，例如新的隐私和安全问题、业务用户的可访问性以及为业务需求选择正确的解决方案。为了利用输入的数据，组织必须解决以下问题：

- 使大数据易于访问。随着数据量的增长，收集和处理数据变得更加困难。组织必须让所有技能水平的数据所有者都可以轻松方便地使用数据。
- 维护质量数据。由于有大量的数据需要维护，组织要花费比以往任何时候都多的时间来清除重复、错误、缺失、冲突和不一致的数据。
- 保持数据安全。随着数据量的增加，隐私和安全问题凸现。在利用大数据之前，组织需要努力实现合规性并实施严格的数据流程。
- 找到合适的工具和平台。处理和分析大数据的新技术一直在发展。组织必须找到合适的技术以在其已建立的生态系统中工作并满足特定需求。通常，正确的解决方案也是一种灵活的解决方案，可以适应未来的基础架构变化。

Bigdata **Exercises**

〖Ex. 1〗 根据 Text A 回答以下问题。

1. What is critical to business growth and innovation in a data-driven world?
2. What is ELT?
3. Why does ELT reduce the time data spends in transit and boost efficiency?
4. What are the three steps ELT works?
5. What are the advantages ELT offers?
6. What are the typical capabilities of ETL tools?
7. What are the primary differences between ETL and ELT?
8. What are the key ways the ETL and ELT approaches to data integration differ in?
9. What can datalakes do?

10. What appears to be the future of data integration? What does it deliver?

〖Ex. 2〗 根据 Text B 回答以下问题。

1. What have many organizations recognized?
2. What does big data analytics describe?
3. What can organizations do with today's technology?
4. Once data is collected and stored, what must it be done?
5. What can dirty data do?
6. What does data mining do?
7. What is Hadoop?
8. What is Spark?
9. What are the big benefits of big data analytics?
10. To capitalize on incoming data, what will organizations have to address?

〖Ex. 3〗 把下列词组翻译成中文。

1. be built into   1. _____
2. computing power   2. _____
3. data transformation   3. _____
4. host service   4. _____
5. storage space   5. _____
6. target database   6. _____
7. total cost of ownership 7. _____
8. batch processing   8. _____
9. cloud storage   9. _____
10. data block   10. _____

〖Ex. 4〗 把下列单词或词组翻译成英文。

1. 流处理   1. _____
2. 供应链   2. _____
3. 虚拟助理   3. _____
4. v. 交付，递送   4. _____
5. adv. 以指数方式   5. _____
6. v. 利用   6. _____
7. n. 规章，规则 adj.规定的 7. _____

8. v. 分配，指派(任务)   8. _____

9. n. 生态系统   9. _____

10. n. 回归   10. _____

〖Ex. 5〗 翻译句子。

1. This technical innovation will save us much time and labour.

2. You can reposition controls using a drag-and-drop operation.

3. It is significantly more compact than any comparable laptop, with no loss in functionality.

4. The new regulation will directly affect us.

5. You need to establish a connection to your target database.

6. The organization created a virtualized infrastructure to run applications.

7. The servers reduce complexity, improve systems management, and increase energy efficiency while driving down total cost of ownership.

8. You need to perform the following steps to create the virtual storage space.

9. Quantum computers may provide much greater computing power than classical computers.

10. This computer provides more computing power and better interactive interfaces.

Bigdata **Reading Material**

Skills Needed by a Data Analyst

1. What Is a Data Analyst?

A data analyst is someone who uses technical skills to analyze data and report insights.

On a typical day, a data analyst might use SQL skills to pull data from a company database, use programming skills to analyze that data, and then use communication skills to report their results to a larger audience.

A data analyst has a fulfilling① job that pays well. Being a data analyst also provides experience that can be beneficial② for stepping into more advanced roles like data scientist.

2. How to Become a Data Analyst?

(1) Learn the technical skills (SQL and some data analysis with Python or R).

(2) Learn the fundamentals of statistics.

---

① fulfilling [fʊlˈfɪlɪŋ] adj. 使人满足的，令人愉快的

② beneficial [ˌbenɪˈfɪʃl] adj. 有益的，有帮助的

(3) Build data analysis projects that showcase[①] your hard and soft skills.

So you've decided you want to be a data analyst. Or maybe your goal is to be a data scientist. In either case, you will need to master some skills to get you where you want to go.

3. What Skills Does a Data Analyst Need?

We'll be focusing on skills and not on tools (like Python, R, SQL, Excel, Tableau, etc.) Our focus will be what you'll need to do as a data analyst, not how you do those things.

Tools—the how—will vary depending on the exact[②] role, the company that hires you, and the industry you end up working in. You can take the data analyst skills from this article and apply them using the tools that you're learning with, or that suit the industry you're looking to break into[③].

3.1 Data cleaning and preparation

Research shows that data cleaning and preparation accounts for around 80% of the work of data professionals. This makes it perhaps the key skill for anyone who is serious about getting a job in data.

Commonly, a data analyst will need to retrieve data from one or more sources and prepare the data so it is ready for numerical and categorical analysis. Data cleaning also involves handling missing and inconsistent[④] data that may affect your analysis.

3.2 Data analysis and exploration

It might sound funny to list "data analysis" in a list of required data analyst skills. But analysis itself is a specific skill that needs to be mastered.

At its core, data analysis means taking a business question or need and turning it into a data question. Then, you'll need to transform and analyze data to extract an answer to that question.

Another form of data analysis is exploration. Data exploration[⑤] is looking to find interesting trends or relationships in the data that could bring value to a business.

Exploration might be guided by an original business question, but it also might be relatively unguided. By looking to find patterns and blips in the data, you may stumble across an opportunity for the business to decrease costs or increase growth!

3.3 Statistical knowledge

A strong foundation in probability and statistics is an important skill for a data analyst. This knowledge will help guide your analysis and exploration and help you understand the data that you're working with.

---

① showcase ['ʃəʊkeɪs]　v. 展示(优点)

② exact [ɪg'zækt]　adj. 确切的；精确的；严谨的

③ break into：进入，闯入

④ inconsistent [ˌɪnkən'sɪstənt]　adj. 不一致的，前后矛盾的，不合逻辑的

⑤ Data exploration：数据探索

Additionally, understanding stats will help you make sure your analysis is valid and will help you avoid common fallacies① and logical② errors.

The exact level of statistical knowledge required will vary depending on the demands of your particular role and the data you're working with. For example, if your company relies on probabilistic③ analysis, you'll need a much more rigorous④ understanding of those areas than you would otherwise.

3.4 Creating data visualizations

Data visualizations make trends and patterns in data easier to understand. Humans are visual creatures, and most people won't be able to get meaningful insights by looking at a giant spreadsheet of numbers. As a data analyst, you'll need to be able to create plots⑤ and charts⑥ to help communicate your data and findings visually.

This means creating clean, visually compelling⑦ charts that will help others understand the data. It also means avoiding things that are either difficult to interpret (like pie charts⑧) or can be misleading⑨ (like manipulating axis values).

Visualizations can also be an important part of data exploration. Sometimes there are things that you can see visually in the data that can hide when you just look at the numbers.

It's very rare to find data role that doesn't require data visualization, and it is deadly an important skill.

3.5 Creating dashboards and/or reports

As a data analyst, you'll need to empower⑩ others within your organization to use data to make key decisions. By building dashboards and reports, you'll be giving others access to important data by removing technical barriers[11].

This might take the form of a simple chart and table with date filters, all the way up to a large dashboard containing hundreds of data points that are interactive and update automatically.

Job requirements can vary a lot from position to position, but almost every data analyst's job is going to involve producing reports on your findings and/or building dashboards to showcase them.

---

① fallacy ['fæləsi] n. 错误，谬误
② logical ['lɒdʒɪkl] adj. 逻辑(上)的，符合逻辑的，推理正确的
③ probabilistic [ˌprɒbəbɪ'lɪstɪk] adj. 概率论的
④ rigorous ['rɪɡərəs] adj. 严格的；严密的；缜密的
⑤ plot [plɒt] n. 图
⑥ chart [tʃɑːt] n. 图表
⑦ compelling [kəm'pelɪŋ] adj. 引人入胜的
⑧ pie chart：饼图
⑨ misleading [ˌmɪs'liːdɪŋ] adj. 误导性的；引入歧途的
⑩ empower [ɪm'paʊə] vt. 授权；准许
11 technical barrier：技术壁垒，技术障碍

3.6 Writing and communication skills[1]

The ability to communicate in multiple formats is a key skill for a data analyst. Writing, speaking, explaining, listening—strong communication skills across all of these areas will help you succeed.

Communication is key in collaborating with your colleagues. For example, in a kickoff meeting with business stakeholders[2], careful listening skills are needed to understand the analyses they require. Similarly, during your project, you may need to be able to explain a complex topic to non-technical teammates[3].

Written communication is also incredibly important—you'll almost certainly need to write up your analysis and recommendations.

Being clear, direct, and easily understood is a skill that will advance your career in data. It may be a "soft" skill, but don't underestimate it—the best analytical skills in the world won't be worth much unless you can explain what they mean and convince your colleagues to act on your findings.

3.7 Domain knowledge

Domain knowledge is understanding things that are specific to the particular industry and company that you work for. For example, if you're working for a company with an online store, you might need to understand the nuances[4] of e-commerce. In contrast, if you're analyzing data about mechanical systems, you might need to understand those systems and how they work.

Domain knowledge changes from industry to industry, so you may find yourself needing to research and learn quickly. No matter where you work, if you don't understand what you're analyzing it's going to be difficult to do it effectively.

This is certainly something that you can learn on the job, but if you know a specific industry or area you'd like to work in, building as much understanding as you can in advance will make you a more attractive job applicant and a more effective employee once you do get the job.

3.8 Problem-solving

As a data analyst, you're going to run up against problems, bugs[5], and roadblocks[6] every day. Being able to problem-solve your way out of them is a key skill.

You might need to research a quirk[7] of some software or coding language that you're

---

① communication skill：沟通技能，沟通技巧

② stakeholder ['steɪkhəʊldə]　n. 股东；利益相关者

③ teammate ['tiːmmeɪt]　n. 队友

④ nuance ['njuːɑːns]　n. 细微差别

⑤ bug [bʌg]　n. 故障

⑥ roadblock ['rəʊdblɒk]　n. 路障，障碍

⑦ quirk [kwɜːk]　n. 怪癖；奇事，巧合

using. Your company might have resource constraints[①] that force you to be innovative in how you approach a problem. The data you're using might be incomplete[②]. Or you might need to perform some "good enough" analysis to meet a looming[③] deadline[④].

Whatever the circumstances, strong problem-solving skills are going to be an incredible asset for any data analyst.

The skills mentioned above are the most important. If you can master them, you are sure to become the best data analyst.

---

① resource constraint: 资源限制，资源约束

② incomplete [ˌɪnkəmˈpliːt] adj. 不完整的，不完全的，不完备的

③ loom [luːm] vi. 迫在眉睫

④ deadline [ˈdedlaɪn] n. 截止时间，最后期限

Unit 6

Bigdata **Text A**

Data Mining

1. What Is Data Mining?

Data mining, also known as knowledge discovery in data (KDD), is the process of uncovering patterns and other valuable information from large data sets. Given the evolution of data warehousing technology and the growth of big data, adoption of data mining techniques has rapidly accelerated over the last couple of decades, and has assisted companies by transforming their raw data into useful knowledge. However, despite the fact that technology continuously evolves to handle data at a large-scale, leaders still face challenges with scalability and automation.

Data mining has improved organizational decision-making through insightful data analyses. The data mining techniques that underpin these analyses can be divided into two main purposes; they can either describe the target dataset or they can predict outcomes through the use of machine learning algorithms. These methods are used to organize and filter data, and to surface the most interesting information, from fraud detection to user behaviors, bottlenecks and even security breaches.

When combined with data analytics and visualization tools, like Apache Spark, delving into the world of data mining has never been easier and extracting relevant insights has never been faster. Advances within artificial intelligence only continue to expedite adoption across industries.

2. Data Mining Process

The data mining process involves a number of steps from data collection to visualization to extract valuable information from large data sets. As mentioned above, data mining techniques are used to generate descriptions and predictions about a target data set. Data scientists describe data through their observations of patterns, associations, and correlations. They also classify and cluster data through classification and regression methods, and identify outliers for use cases, like spam detection.

Data mining usually consists of the following four main steps:

2.1　Set the business objectives

This can be the hardest part of the data mining process, and many organizations spend too little time on this important step. Data scientists and business stakeholders need to work together to define the business problem, which helps inform the data questions and parameters for a given project. Analysts may also need to do additional research to understand the business context appropriately.

2.2　Data preparation

Once the scope of the problem is defined, it is easier for data scientists to identify which set of data will help answer the pertinent questions to the business. Once they collect the relevant data, the data will be cleaned, that is removing any noise, such as duplicates, missing values, and outliers. Depending on the dataset, an additional step may be taken to reduce the number of dimensions as too many features can slow down any subsequent computation. Data scientists will look to retain the most important predictors to ensure optimal accuracy within any models.

2.3　Model building and pattern mining

Depending on the type of analysis, data scientists may investigate any interesting data relationships, such as sequential patterns, association rules, or correlations. While high frequency patterns have broader applications, sometimes the deviations in the data can be more interesting, highlighting areas of potential fraud.

Deep learning algorithms may also be applied to classify or cluster a data set depending on the available data. If the input data is labelled (i.e. supervised learning), a classification model may be used to categorize data, or alternatively, a regression may be applied to predict the likelihood of a particular assignment. If the dataset isn't labelled (i.e. unsupervised learning), the individual data points in the training set are compared with one another to discover underlying similarities, and cluster them based on those characteristics.

2.4　Evaluation of results and implementation of knowledge

Once the data is aggregated, the results need to be evaluated and interpreted. When finalizing results, they should be valid, novel, useful, and understandable. When this criteria is met, organizations can use this knowledge to implement new strategies to achieve their intended objectives.

3.　Data Mining Techniques

Data mining works by using various algorithms and techniques to turn large volumes of data into useful information. Here are some of the most common ones.

3.1　Association rules

An association rule is a rule-based method for finding relationships between variables in a given dataset. These methods are frequently used for market basket analysis, which allows companies to better understand relationships between different products. Understanding consumption habits of customers enables businesses to develop better cross-selling strategies and recommendation engines.

3.2 Neural networks

Primarily leveraged for deep learning algorithms, neural networks process training data by mimicking the interconnectivity of the human brain through layers of nodes. Each node is made up of inputs, weights, a bias (or threshold) and an output. If that output value exceeds a given threshold, it "fires" or "activates" the node and passes data to the next layer in the network. Neural networks learn this mapping function through supervised learning and adjust based on the loss function through the process of gradient descent. When the cost function is at or near zero, we can be confident in the model's accuracy to yield the correct answer.

3.3 Decision tree

This data mining technique uses classification or regression methods to classify or predict potential outcomes based on a set of decisions. As the name suggests, it uses a tree-like visualization to represent the potential outcomes of these decisions.

3.4 K-nearest neighbor (KNN)

K-nearest neighbor, also known as the KNN algorithm, is a non-parametric algorithm that classifies data points based on their proximity and association to other available data. This algorithm assumes that similar data points can be found near each other. As a result, it seeks to calculate the distance between data points, usually through Euclidean distance, and then it assigns a category based on the most frequent category or average.

4. Data Mining Applications

Data mining techniques are widely adopted among business intelligence and data analytics teams. They help them extract knowledge for their organization and industry. The following are some data mining use cases:

4.1 Sales and marketing

Companies collect a massive amount of data about their customers and prospects. By observing consumer demographics and online user behavior, companies can use data to optimize their marketing campaigns, improve segmentation, cross-sell offers, and customer loyalty programs, and yield higher ROI on marketing efforts. Predictive analyses can also help teams to set expectations with their stakeholders, and provide yield estimates from any increases or decreases in marketing investment.

4.2 Education

Educational institutions have started to collect data to understand their student populations as well as which environments are conducive to success. As courses continue to transfer to online platforms, they can use a variety of dimensions and metrics to observe and evaluate performance, such as keystroke, student profiles, classes, universities, time spent, etc.

4.3　Operational optimization

Process mining leverages data mining techniques to reduce costs across operational functions, and enable organizations to run more efficiently. This practice has helped to identify costly bottlenecks and improve decision-making among business leaders.

4.4　Fraud detection

While frequently occurring patterns in data can provide teams with valuable insight, observing data anomalies is also beneficial to assisting companies in detecting fraud. While this is a well-known use case within banking and other financial institutions, SaaS-based companies have also started to adopt these practices to eliminate fake user accounts from their datasets.

New Words

evolution	[ˌiːvə'luːʃn]	n. 进化；演变
adoption	[ə'dɒpʃn]	n. 采用
rapidly	['ræpɪdlɪ]	adv. 快速地，迅速地
accelerate	[ək'seləreɪt]	v. (使)加速，加快
knowledge	['nɒlɪdʒ]	n. 知识
continuously	[kən'tɪnjʊəslɪ]	adv. 连续不断地
insightful	['ɪnsaɪtfʊl]	adj. 富有洞察力的，有深刻见解的
bottleneck	['bɒtlnek]	n. 瓶颈
description	[dɪ'skrɪpʃn]	n. 描述，说明
inform	[ɪn'fɔːm]	v. 了解；通知；对……有影响
appropriately	[ə'prəʊprɪətli]	adv. 适当地
pertinent	['pɜːtɪnənt]	adj. 有关的，相干的
dimension	[daɪ'menʃn]	n. 维度；尺寸；范围；规模
subsequent	['sʌbsɪkwənt]	adj. 后来的；随后的
optimal	['ɒptɪməl]	adj. 最佳的，最优的
accuracy	['ækjərəsi]	n. 精确(性)，准确(性)
investigate	[ɪn'vestɪgeɪt]	v. 调查；研究
sequential	[sɪ'kwenʃl]	adj. 按次序的，序列的
highlight	['haɪlaɪt]	v. 突出，强调
potential	[pə'tenʃl]	adj. 潜在的 n. 潜力；可能性

likelihood	['laɪklɪhʊd]	n. 可能，可能性
similarity	[ˌsɪmə'lærəti]	n. 相似性，类似点
novel	['nɒvl]	adj. 新颖的
understandable	[ˌʌndə'stændəbl]	adj. 能懂的，可理解的
mimic	['mɪmɪk]	v. 模仿，模拟
interconnectivity	[ˌɪntəkənek'tɪvɪti]	n. 互联性，相互连接性
weight	[weɪt]	n. 权重
bias	['baɪəs]	n. 偏差
threshold	['θreʃhəʊld]	n. 阈值
proximity	[prɒk'sɪməti]	n. 接近度，距离
assume	[ə'sjuːm]	v. 假设，认为
average	['ævərɪdʒ]	n. 平均数；平均水平
		v. 平均为；算出……的平均数
		adj. 平均的
prospect	['prɒspekt]	n. 前景；期望
loyalty	['lɔɪəltɪ]	n. 忠诚，忠实
expectation	[ˌekspek'teɪʃn]	n. 期待，希望
yield	[jiːld]	n. 收益；产量
estimate	['estɪmət]	n. 估计，预测
	['estɪmeɪt]	v. 估计，估算
keystroke	['kiːstrəʊk]	n. 按键

Phrases

large data set	大数据集
be divided into …	被分为……
user behavior	用户行为
security breach	安全漏洞
delving into	深入研究
use case	用例
spam detection	垃圾邮件检测
slow down	(使)慢下来
model building	模型构建，建模
pattern mining	模式挖掘
sequential pattern	序列模式，顺序模式
association rule	关联规则

classification model	分类模型
intended objective	预期目标
turn ... into ...	把……转化为……
rule-based method	基于规则的方法
market basket	购物篮
consumption habit	消费习惯
recommendation engine	推荐引擎
neural network	神经网络
deep learning	深度学习
loss function	损失函数，损耗函数
cost function	成本函数
decision tree	决策树
non-parametric algorithm	非参数算法
financial institution	金融机构

Abbreviations

KDD (Knowledge Discovery in Data)	数据中的知识发现
KNN (K-Nearest Neighbor)	K 最近邻
ROI (Return On Investment)	投资回报率，投资利润

 ## Analysis of Difficult Sentences

[1] When combined with data analytics and visualization tools, like Apache Spark, delving into the world of data mining has never been easier and extracting relevant insights has never been faster.

本句中，and 连接了两个并列句，其中 delving into the world of data mining 和 extracting relevant insights 是动名词短语，作主语；When combined with data analytics and visualization tools, like Apache Spark 做条件状语。

[2] Data scientists and business stakeholders need to work together to define the business problem, which helps inform the data questions and parameters for a given project.

本句中，which helps inform the data questions and parameters for a given project 是非限定性定语从句，对前面整个句子进行补充说明。which 指前面的整个句子。

[3] Depending on the dataset, an additional step may be taken to reduce the number of dimensions as too many features can slow down any subsequent computation.

本句中，Depending on the dataset 是条件状语，修饰谓语 may be taken。to reduce the number of dimensions 是动词不定式短语，作目的状语，也修饰谓语。as too many features can slow down any subsequent computation 是原因状语从句，修饰谓语。

[4] Primarily leveraged for deep learning algorithms, neural networks process training data by mimicking the interconnectivity of the human brain through layers of nodes.

　　本句中，Primarily leveraged for deep learning algorithms 是过去分词短语，作 neural networks 的同位语，对其进行补充说明。by mimicking the interconnectivity of the human brain through layers of nodes 是介词短语，作方式状语，修饰谓语 process。

 参考译文

数 据 挖 掘

1．什么是数据挖掘？

　　数据挖掘也称为数据知识发现(KDD)，是指从大型数据集中发现模式和其他有价值信息的过程。鉴于数据仓库技术的发展和大数据的增长，数据挖掘技术的应用在过去几十年中已经迅速加速，并通过将原始数据转化为有用的知识来帮助公司。然而，尽管该技术不断发展以大规模处理数据，但领导者仍然面临可扩展性和自动化方面的挑战。

　　数据挖掘通过富有洞察力的数据分析改进了组织决策。支持这些分析的数据挖掘技术有两个主要目的：它们可以描述目标数据集，也可以通过使用机器学习算法预测结果。这些方法可用于组织和过滤数据，并呈现从欺诈检测到用户行为、瓶颈，甚至安全漏洞等有趣的信息。

　　当与数据分析和可视化工具(如 ApacheSpark)结合使用时，深入挖掘数据前所未有地简单，提取相关见解也前所未有地快速。人工智能领域的进步会继续加快各个行业采用这些技术。

2．数据挖掘过程

　　数据挖掘过程涉及从数据收集到可视化以从大数据集中提取有价值信息的多个步骤。如上所述，数据挖掘技术用于生成关于目标数据集的描述和预测。数据科学家通过对模式、关联和相关性的观察来描述数据。他们还通过分类和回归方法对数据进行分类和聚类，并识别用例的异常值，例如垃圾邮件检测。

　　数据挖掘通常包括以下四个主要步骤：

2.1　设定业务目标

　　设定业务目标可能是数据挖掘过程中最困难的部分，许多组织在这一重要步骤上花费的时间很少。数据科学家和业务利益相关者需要共同努力定义业务问题，这有助于了解项目的数据问题和参数。分析师可能还需要进行额外的研究来适当地了解业务背景。

2.2 数据准备

一旦确定了问题的范围，数据科学家就更容易确定哪组数据将有助于解决与业务相关的问题。一旦他们收集了相关数据，数据将被清理，即去除任何噪声(例如重复、缺失值和异常值)。根据数据集的情况，可能会采取额外的步骤来减少维数，因为特性太多会降低后续计算的速度。数据科学家将保留最重要的预测因子，以确保模型的准确性最佳。

2.3 模型构建与模式挖掘

根据分析的类型，数据科学家可能会调查有趣的数据关系，例如顺序模式、关联规则或相关性。虽然高频模式应用更广泛，但有时数据中的偏差可能更有趣，突出了潜在的欺诈领域。

深度学习算法也可用于根据可用数据对数据集进行分类或聚类。如果输入数据被标记(即监督学习)，则可以使用分类模型对数据进行分类，也可以使用回归来预测特定分配的可能性。如果数据集没有标记(即无监督学习)，则将训练集中的各个数据点相互比较以发现潜在的相似性，并根据这些特征对它们进行聚类。

2.4 结果评价与知识实施

汇总数据后，需要对结果进行评估和解释。在最终确定结果时，数据应该是有效的、新颖的、有用的和易于理解的。当满足此标准时，组织可以使用这些知识来实施新策略实现其预期目标。

3. 数据挖掘技术

数据挖掘的工作原理是使用各种算法和技术将大量数据转化为有用的信息。以下是一些最常见的技术。

3.1 关联规则

关联规则是一种基于规则的方法，用于查找给定数据集中变量之间的关系。这些方法经常用于购物篮分析，使公司能够更好地了解不同产品之间的关系。了解客户的消费习惯可帮助企业开发更好的交叉销售策略和推荐引擎。

3.2 神经网络

神经网络主要用于深度学习算法，它通过节点层模仿人脑的互连性来处理训练数据。每个节点由输入、权重、偏差(或阈值)和输出组成。如果输出值超过给定的阈值，则会"触发"或"激活"节点并将数据传递到网络中的下一层。神经网络通过监督学习来学习该映射函数，并根据梯度下降过程中的损失函数进行调整。当成本函数为零或接近零时，我们可以确信模型的准确性，从而得出正确的答案。

3.3 决策树

决策树使用分类或回归方法根据一组决策对潜在结果进行分类或预测。顾名思义，它使用树状可视化形式来表示这些决策的潜在结果。

3.4 K最近邻(KNN)

K最近邻(KNN)是一种非参数算法，它根据数据点与其他可用数据的接近度和关联性对数据进行分类。该算法假设可以在彼此附近找到相似的数据点。因此，它通常通过欧几里得距离寻求计算数据点之间的距离，然后根据最常见的类别或平均值分配一个类别。

4. 数据挖掘应用

数据挖掘技术被商业智能和数据分析团队广泛采用。这些技术帮助他们为组织和行业获取知识。以下介绍一些数据挖掘用例。

4.1 销售和营销

公司收集有关客户和潜在客户的大量数据。通过观察消费者人口统计数据和在线用户行为，可以使用数据来优化营销活动，改进细分、交叉销售和客户忠诚度计划，并在营销工作中产生更高的投资回报率。预测分析还可以帮助团队与其利益相关者设定期望，并根据营销投资的增加或减少估计收益。

4.2 教育

教育机构通过收集数据来了解学生群体以及哪些环境有利于成功。随着课程不断转移到在线平台，他们可以从各个方面使用各种指标来观察和评估绩效，例如点击次数、学生档案、课程、学校、花费的时间等。

4.3 运营优化

流程挖掘利用数据挖掘技术来降低运营成本，并使组织能更高效运行。这种做法有助于识别成本高昂的瓶颈并优化企业领导者的决策。

4.4 欺诈检测

虽然数据中频繁出现的模式可以为团队提供宝贵的见解，但观察数据异常也有利于帮助公司发现欺诈行为。虽然这是银行和其他金融机构的众所周知的用例，但基于 SaaS 的公司也开始采用这些做法来从其数据集中消除虚假用户账户。

Bigdata **Text B**

Best Data Mining Tools

1. Xplenty

Xplenty provides a platform that has functionalities to integrate, process, and prepare data for analytics. Businesses will be able to make most of the opportunities offered by big data with the help of Xplenty and that too without investing in related personnel, hardware, and software. It is a complete toolkit for building data pipelines.

You will be able to implement complex data preparation functions through rich expression language. It has an intuitive interface to implement ETL, ELT, or a replication solution. You will be able to orchestrate and schedule pipelines through a workflow engine.

Xplenty has the following features:

• Xplenty is the data integration platform for all. It offers the no-code and low-code options.

• An API component will provide advanced customization and flexibility.

• It has functionalities to transfer and transform data between databases and data warehouses.

• It provides support through email, chat, phone, and an online meeting.

2. Rapid Miner

Rapid Miner is one of the best predictive analysis system developed by the company. It is written in Java programming language. It provides an integrated environment for deep learning, text mining, machine learning and predictive analysis.

The tool can be used for over a vast range of applications including for business applications, commercial applications, training, education, research, application development as well as machine learning.

Rapid Miner offers the server as both on premise and in public/private cloud infrastructures. It has a client/server model as its base. Rapid Miner comes with template based frameworks that enable speedy delivery with reduced number of errors (which are quite commonly expected in manual code writing process).

Rapid Miner constitutes of three modules, namely:

• Rapid Miner Studio: This module is for workflow design, prototyping, validation etc.

• Rapid Miner Server: This is to operate predictive data models created in studio.

• Rapid Miner Radoop: It execute processes directly in the Hadoop cluster to simplify predictive analysis.

3. Orange

Orange is a perfect software suite for machine learning and data mining. It best aids the data visualization and is a component based software. It is written in Python computing language.

As it is a component-based software, the components of Orange are called "widgets". These widgets range from data visualization and pre-processing to an evaluation of algorithms and predictive modeling.

Widgets offer major functionalities like:

• Showing data table and allowing to select features.

• Reading the data.

• Training predictors and comparing learning algorithms.

• Visualizing data elements etc.

Additionally, Orange brings a more interactive and fun vibe to the dull analytic tools. It is quite interesting to operate.

Data coming to Orange gets quickly formatted to the desired pattern and it can be easily moved where needed by simply moving the widgets. Users are quite fascinated by Orange. Orange allows users to make smarter decisions in short time by quickly comparing and analyzing the data (see Figure 6-1).

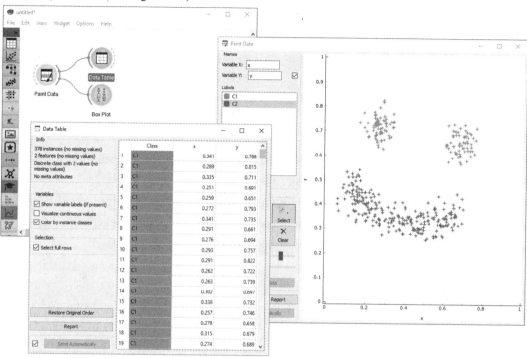

Figure 6-1   Orange: Paint a Two-dimensional Data Set

4. Weka

Weka, also known as Waikato Environment, is a machine learning software developed at the University of Waikato in New Zealand. It is best suited for data analysis and predictive modeling. It contains algorithms and visualization tools that support machine learning.

Weka has a GUI that facilitates easy access to all its features. It is written in Java programming language. It supports major data mining tasks including data mining, processing, visualization, regression etc. It works on the assumption that data is available in the form of a flat file. It can provide access to SQL databases through database connectivity and can further process the data/results returned by the query.

5. KNIME

KNIME is the best integration platform for data analytics and reporting developed by KNIME.com AG. It operates on the concept of the modular data pipeline. KNIME constitutes of various machine learning and data mining components embedded together.

KNIME has been used widely for pharmaceutical research. In addition, it performs excellently for customer data analysis, financial data analysis, and business intelligence.

KNIME has some brilliant features like quick deployment and scaling efficiency. Users get familiar with KNIME in quite lesser time and it has made predictive analysis accessible to even naive users. KNIME utilizes the assembly of nodes to pre-process the data for analytics and visualization.

6. Sisense

Sisense is extremely useful and best suited BI software when it comes to reporting purposes within the organization. It is developed by the company of the same name "Sisense". It has a brilliant capability to handle and process data for the small scale/large scale organizations.

It allows combining data from various sources to build a common repository and further, refines data to generate rich reports that get shared across departments.

Sisense generates reports which are highly visual. It is specially designed for users that are non-technical. It allows drag and drop facility as well as widgets.

Different widgets can be selected to generate the reports in form of pie charts, line charts, bar graphs etc. based on the purpose of an organization. Reports can be further drilled down by simply clicking to check details and comprehensive data.

7. SSDT (SQL Server Data Tools)

SSDT is a universal, declarative model that expands all the phases of database development in the Visual Studio IDE. Developers use SSDT transact—a design capability of SQL—to build, maintain, debug and refactor databases.

A user can work directly with a database or can work directly with a connected database, thus, providing on or off-premise facility. SSDT provides Table Designer to create new tables as well as edit tables in direct databases as well as connected databases.

8. Apache Mahout

Apache Mahout is a project developed by Apache Foundation that serves the primary purpose of creating machine learning algorithms. It focuses mainly on data clustering, classification, and collaborative filtering.

Mahout is written in Java and includes Java libraries to perform mathematical operations like linear algebra and statistics. Mahout is growing continuously as the algorithms implemented inside Apache Mahout are continuously growing.

To key up, Mahout has following major features:
- Extensible programming environment.
- Pre-made algorithms.
- Math experimentation environment.
- GPU computes for performance improvement.

9. Oracle Data Mining

A component of Oracle Advance Analytics, Oracle data mining software provides excellent data mining algorithms for data classification, prediction, regression and specialized analytics that enables analysts to analyze insights, make better predictions, target best customers, identify cross-selling opportunities and detect fraud.

The algorithms designed inside Oracle Data Mining leverage the potential strengths of Oracle database. The data mining feature of SQL can dig data out of database tables, views, and schemas.

The GUI of Oracle Data Mining is an extended version of Oracle SQL Developer. It provides a facility of direct "drag and drop" of data inside the database to users thus giving better insight.

10. IBM Cognos

IBM Cognos is an intelligence suite owned by IBM for reporting and data analysis, score carding etc. It consists of sub-components that meet specific organizational requirements Cognos Connection, Query Studio, Report Studio, Analysis Studio, Event studio and Workspace Advance.

- Cognos Connection: A web portal to gather and summarize data in scoreboard/reports.
- Query Studio: To contain queries to format data and create diagrams.
- Report Studio: To generate management reports.
- Analysis Studio: To process large data volumes, understand and identify trends.
- Event Studio: Notification module to keep in sync with events.
- Workspace Advanced: User-friendly interface to create personalized and user-friendly documents.

11. IBM SPSS Modeler

IBM SPSS is a software suite owned by IBM that is used for data mining and text analytics to build predictive models. It was originally produced by SPSS Inc. and later on acquired by IBM.

SPSS Modeler has a visual interface that allows users to work with data mining algorithms without the need of programming. It eliminates the unnecessary complexities faced during data transformations and to make easy to use predictive models.

12. SAS

Statistical Analysis System (SAS) is a product of SAS Institute developed for analytics and data management. SAS can mine data, alter it, manage data from different sources and perform statistical analysis. It provides a GUI for non-technical users.

SAS data miner enables users to analyze big data and derives accurate insight to make timely decisions. SAS has a distributed memory processing architecture which is highly scalable. It is well suited for data mining, text mining and optimization.

New Words

toolkit	['tuːlkɪt]	n. 工具包，工具箱
pipeline	['paɪplaɪn]	n. 管道；渠道
expression	[ɪk'spreʃn]	n. 表达式
intuitive	[ɪn'tjuːɪtɪv]	adj. 直观的；直觉的
template	['templeɪt]	n. 样板；模板
prototype	['prəʊtətaɪp]	n. 原型
suite	[swiːt]	n. (软件的)套件
widget	['wɪdʒɪt]	n. (窗口)小部件
vibe	[vaɪb]	n. 气氛
dull	[dʌl]	adj. 枯燥无味的
fascinate	['fæsɪneɪt]	v. 使着迷
		v. 入迷
major	['meɪdʒə]	adj. 主要的，重要的
connectivity	[ˌkɒnek'tɪvɪti]	n. 连通性
modular	['mɒdjələ]	adj. 模块化的
pharmaceutical	[ˌfɑːmə'suːtɪkl]	adj. 制药的，配药的
		n. 药物
brilliant	['brɪljənt]	adj. 出色的；非常好的；成功的
assembly	[ə'sembli]	n. 组装，装配
refine	[rɪ'faɪn]	v. 提炼；改善；细化
declarative	[dɪ'klærətɪv]	adj. 声明的，宣言的，陈述的
refactor	[ri'fæktə]	n. 重构
experimentation	[ɪkˌsperɪmen'teɪʃn]	n. 实验，试验
scoreboard	['skɔːbɔːd]	n. 记分板
sync	[sɪŋk]	n. 同时，同步(synchronization的缩略词)
		v. 使同步(synchronize的缩略词)
user-friendly	[ˌjuːzə'frendli]	adj. 用户友好的

Phrases

workflow engine	工作流引擎
online meeting	在线会议
predictive analysis system	预测分析系统

integrated environment	集成环境
text mining	文本挖掘
client/server model	客户机/服务器模式
data table	数据表
flat file	平面文件
constitute of	构成
bar graph	条形图
linear algebra	线性代数
web portal	门户网站
predictive model	预测模型
distributed memory processing architecture	分布式内存处理架构

Abbreviations

GPU (Graphics Processing Unit)	图形处理器
SPSS (Statistical Product and Service Solutions)	统计产品与服务解决方案
SAS (Statistical Analysis System)	统计分析系统

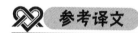 参考译文

最佳数据挖掘工具

1. Xplenty

Xplenty 是一个具有集成、处理和准备数据以进行分析功能的平台。在 Xplenty 的帮助下，企业将能够充分利用大数据提供的大部分机会，而无需投资人员、硬件和软件。它是构建数据管道的完整工具包。

用户能够通过丰富的表达式语言实现复杂的数据准备功能。Xplenty 有一个直观的界面来实现 ETL、ELT 或复制解决方案。用户能够通过工作流引擎编排和安排管道。

Xplenty 具有以下特点：

- Xplenty 是适用于所有人的数据集成平台。它提供无代码和低代码选项。
- API 组件将提供高级定制和灵活性。
- 具有在数据库和数据仓库之间传输和转换数据的功能。
- 通过电子邮件、聊天、电话和在线会议提供支持。

2. Rapid Miner

Rapid Miner 是最好的预测分析系统之一。Rapid Miner 是用 Java 编程语言编写的。它为深度学习、文本挖掘、机器学习和预测分析提供了一个集成环境。

该工具可用的应用程序十分广泛，包括企业应用程序、商业应用程序、培训、教育、研究、应用程序开发以及机器学习。

Rapid Miner 将服务器作为内部部署和公共/私有云基础设施提供。它以客户端/服务器模型为基础。它具有基于模板的框架，可以在减少错误数量的情况下实现快速交付(这在手动代码编写过程中很常见)。

Rapid Miner 由三个模块构成，分别是：

- Rapid Miner Studio：该模块用于工作流设计、原型设计、验证等。
- Rapid Miner Server：该模块用于操作在工作室中创建的预测数据模型。
- Rapid Miner Radoop：该模块直接在 Hadoop 集群中执行流程以简化预测分析。

3．Orange

Orange 是用于机器学习和数据挖掘的完美软件套件。它有助于数据可视化，是一个基于组件的软件。它是用 Python 语言编写的。

由于 Orange 是一个基于组件的软件，其组件被称为"小部件"。这些小部件的范围包括从数据可视化和预处理到算法评估和预测建模。

小部件提供的主要功能如下：

- 显示数据表并允许选择特征。
- 读取数据。
- 训练预测器并比较学习算法。
- 可视化数据元素等。

此外，Orange 为枯燥的分析工具带来了更具交互性和趣味性的氛围，操作起来颇为有趣。

进入 Orange 的数据会快速格式化为所需的模式，只需移动小部件即可轻松移动到需要的地方。用户对 Orange 非常感兴趣。Orange 允许用户通过快速比较和分析数据，在短时间内做出更明智的决策(图 6-1)。

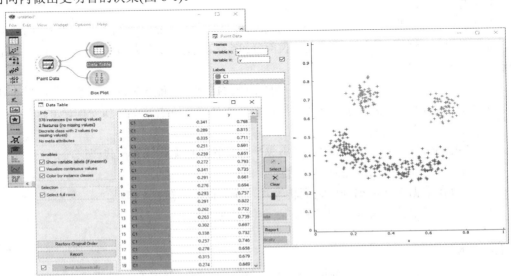

图 6-1　Orange：二维数据集的绘制

4．Weka

Weka 也称为 Waikato 环境，是由新西兰 Waikato 大学开发的机器学习软件。它最适合数据分析和预测建模。Weka 包含支持机器学习的算法和可视化工具。

Weka 有一个 GUI，可以方便地访问其所有功能。Weka 是用 Java 编程语言编写的。它支持主要的数据挖掘任务，包括数据挖掘、处理、可视化、回归等。它假设可用的数据为平面文件形式。它可以通过连接数据库对 SQL 数据库进行访问，并进一步处理查询返回的数据/结果。

5．KNIME

KNIME 是由 KNIME.com AG 开发的最佳的数据分析和报告集成平台。它基于模块化数据管道的概念运行。KNIME 由嵌入在一起的各种机器学习和数据挖掘组件组成。

KNIME 已广泛用于药物研究。此外，它也在客户数据分析、财务数据分析和商业智能方面表现出色。

KNIME 具有一些出色的功能，例如快速部署和扩展效率。用户可在更短的时间内熟悉 KNIME，即使新手也可以用它进行预测分析。KNIME 利用节点组合来预处理数据以进行分析和可视化。

6．Sisense

在企业内提交报告时，Sisense 是非常有用且适合的商务智能软件。它由同名的"Sisense"公司开发，具有为小型/大型组织掌握和处理数据的卓越能力。

Sisense 允许组合不同来源的数据来构建公共存储库，并进一步细化数据以生成可以跨部门共享的丰富报告。

Sisense 可生成高度可视化的报告。它是专门为非技术用户设计的。它允许使用拖放功能部件以及小部件。

可以根据企业报告的目的选择不同的小部件来生成饼图、折线图、条形图等形式。只需单击即可查看详细信息和综合数据，从而进一步深入了解分析报告。

7．SSDT(SQL Server 数据工具)

SSDT 是一种通用的声明性模型，它扩展了 Visual Studio IDE 中数据库开发的所有阶段。开发人员使用 SSDT 事务——SQL 的一种设计功能——来构建、维护、调试和重构数据库。

用户可以直接使用数据库或可以直接连接数据库，从而提供内部或外部设施。SSDT 提供表设计器来创建新表以及编辑直接数据库和连接数据库中的表。

8. Apache Mahout

Apache Mahout 是由 Apache Foundation 开发的项目，其主要目的是创建机器学习算法。

它主要关注数据聚类、分类和协同过滤。

Mahout 是用 Java 编写的，并包含 Java 库来执行线性代数和统计等数学运算。Mahout 的使用在不断增长，因为在 Apache Mahout 中实现的算法也在不断增长。

Mahout 具有以下主要特点：

- 可扩展的编程环境。
- 预先制作的算法。
- 数学实验环境。
- GPU 计算以提高性能。

9．Oracle 数据挖掘

作为 Oracle Advance Analytics 的一个组件，Oracle 数据挖掘软件为数据分类、预测、回归和专业分析提供了出色的数据挖掘算法，使分析师能够分析见解、做出更好的预测、瞄准最佳客户、识别交叉销售机会和检测欺诈。

Oracle Data Mining 内部设计的算法利用了 Oracle 数据库的潜在优势。SQL 的数据挖掘功能可以从数据库表、视图和模式中挖掘数据。

Oracle Data Mining 的 GUI 是 Oracle SQL Developer 的扩展版本。它为用户提供了直接"拖放"数据库内数据的便利，从而提供了更好的洞察力。

10．IBM Cognos

IBM Cognos 是 IBM 拥有的用于报告和数据分析、记分卡等的智能套件。它由满足特定组织要求的子组件 Cognos Connection、Query Studio、Report Studio、Analysis Studio、Event Studio 和 Workspace Advance 组成。

- Cognos Connection：一个网络门户，用于收集和汇总记分板/报告中的数据。
- Query Studio：包含用于格式化数据和创建图表的查询。
- Report Studio：生成管理报告。
- Analysis Studio：处理大量数据，了解和识别趋势。
- Event Studio：通知模块与事件保持同步。
- Workspace Advanced：用户友好的界面，用于创建个性化和用户友好的文档。

11．IBM SPSS Modeler

IBM SPSS 是 IBM 拥有的软件套件，用于数据挖掘和文本分析以构建预测模型。它最初由 SPSS Inc.生产，后来被 IBM 收购。

SPSS Modeler 具有可视化界面，使用户无须编程即可使用数据挖掘算法。它消除了数据转换过程中不必要的复杂性，使预测模型易于使用。

12. SAS

统计分析系统(SAS)是 SAS 研究所为分析和数据管理而开发的产品。SAS 可以挖掘数据、修改数据、管理不同来源的数据并执行统计分析。它为非技术用户提供图形用户界面。

SAS 数据挖掘器使用户能够分析大数据并获得准确的洞察力以便及时做出决策。SAS 具有高度可扩展的分布式内存处理架构。它非常适用于数据挖掘、文本挖掘和优化。

Bigdata **Exercises**

〖Ex. 1〗 根据 Text A 回答以下问题。

1. What is data mining?

2. What does data mining usually consist of?

3. What may data scientists investigate?

4. Once the data is aggregated, what do the results need to be done? When finalizing results, what should they be?

5. How does data mining work?

6. What is an association rule?

7. What is K-nearest neighbor?

8. What are some data mining use cases?

9. By observing consumer demographics and online user behavior, what can companies do?

10. What does process mining leverage data mining techniques to do?

〖Ex. 2〗 根据 Text B 回答以下问题。

1. What does Xplenty do?

2. What does Rapid Miner do?

3. What does Rapid Miner constitute of?

4. What major functionalities do widgets offer?

5. What is KNIME?

6. What is SSDT?

7. What major features does Mahout have?

8. What does Oracle data mining software do?

9. What does IBM Cognos consist of?

10. What can SAS do?

〖Ex. 3〗 把下列词组翻译成中文。

1. classification model   1. _____

_____   _____

2. decision tree   2. _____

3. deep learning   3. _____

4. market basket   4. _____

5. neural network   5. _____

6. pattern mining   6. _____

7. recommendation engine   7. _____

8. security breach   8. _____

9. spam detection   9. _____

10. use case   10. _____

〖Ex. 4〗 把下列单词或词组翻译成英文。

1. 条形图   1. _____

2. 数据表   2. _____

3. 预测模型   3. _____

4. 文本挖掘   4. _____

5. n. 精确(性)，准确(性)   5. _____

6. n. 偏差   6. _____

7. n. 估计，预测 vt. 估计，估算   7. _____

8. n. 互联性，相互连接性   8. _____

9. adj. 最佳的，最优的   9. _____

10. n. 组装，装配   10. _____

〖Ex. 5〗 翻译句子。

1. The keynote speech is about the evolution of the computer.

2. Growth will accelerate to 2.9 per cent next year.

3. Practices of many methods and theories on large data set also need deeper research.

4. If there is a security breach in a small area, it can permeate the entire system.

5. This may significantly impact the execution time of the rule when executed on a large data set.

6. You can use this information to fix a security breach.

7. The use case technology has been used widely in most software development and design methods.

8. Content-based spam detection is one of the most widely used techniques.

9. Sequence pattern mining is developed based on association rules mining.

10. Recommendation engine, which is configured on the distributed development framework, analyzes users' history tags and recommends videos users may like.

Bigdata **Reading Material**

Data Lake

1. Definition of Data Lake

A data lake[①] is a storage repository that holds a vast amount of raw data in its native format[②] until it is needed. While a hierarchical[③] data warehouse stores data in files or folders[④], a data lake uses a flat architecture to store data. Each data element in a lake is assigned a unique identifier[⑤] and tagged with a set of extended metadata tags[⑥]. When a business question arises, the data lake can be queried for relevant data, and that smaller set of data can then be analyzed to help answer the question.

The term data lake is often associated with Hadoop-oriented object storage. In such a scenario, an organization's data is first loaded into the Hadoop platform, and then business analytics and data mining tools are applied to the data that resides on Hadoop's cluster nodes[⑦] of commodity computers.

Like big data, the term data lake is sometimes disparaged[⑧] as being simply a marketing label for a product that supports Hadoop. Increasingly, however, the term is being used to describe any large data pool[⑨] in which the schema and data requirements are not defined until the data is queried.

The term describes a data storage strategy, not a specific technology, although it is frequently used in conjunction with[⑩] a specific technology (Hadoop). The same can be said of the term data warehouse, which despite often referring to a specific technology (relational database), actually describes a broad data management strategy.

---

① data lake：数据湖

② native format：原生格式

③ hierarchical [ˌhaɪəˈrɑːkɪkl]　adj. 分层的，层级的

④ folder [ˈfəʊldə]　n. 文件夹

⑤ identifier [aɪˈdentɪfaɪə]　n. 标识符，识别符

⑥ metadata tag：元数据标签

⑦ cluster node：集群节点

⑧ disparage [dɪˈspærɪdʒ]　vt. 贬低；批评；非难

⑨ data pool：数据池

⑩ in conjunction with…　与……协作

2. Data Lake Architecture

The physical architecture of a data lake may vary, as data lake is a strategy that can be applied to multiple technologies. For example, the physical architecture of a data lake using Hadoop might differ from that of data lake using Amazon Simple Storage Service① (Amazon S3).

However, there are three main principles② that distinguish③ a data lake from other big data storage methods and make up the basic architecture of a data lake. They are:

• No data is turned away. All data is loaded in from various source systems and retained.

• Data is stored in an untransformed or nearly untransformed state④, as it was received from the source.

• Data is transformed and fit into a schema based on analysis requirements.

Although data is largely unstructured and not geared toward answering any specific question, it should still be organized in some manner so that doing this in the future is possible. Whatever technology is used to deploy⑤ an organization's data lake, a few features should be included to ensure that the data lake is functional and healthy and that the large repository of unstructured data doesn't go to waste. These include:

• A taxonomy⑥ of data classifications, which can include data type, content, usage scenarios and groups of possible users.

• A file hierarchy with naming conventions⑦.

• Data profiling tools⑧ to provide insight for classifying data objects and addressing data quality issues.

• Standardized data access process to keep track of what members of an organization are accessing data.

• A searchable⑨ data catalog.

• Data protections including data masking, data encryption and automated monitoring to generate alerts when data is accessed by unauthorized parties.

• Data awareness among employees, which includes an understanding of proper data

---

① Amazon Simple Storage Service：亚马逊简单存储服务
② principle ['prɪnsəpl]　n. 原则，法则
③ distinguish [dɪ'stɪŋgwɪʃ]　v. 区分，使有别于
④ untransformed state：未转换状态
⑤ deploy [dɪ'plɔɪ]　v. 部署
⑥ taxonomy [tæk'sɒnəmi]　n. 分类学，分类系统
⑦ convention [kən'venʃn]　n. 惯例；约定
⑧ data profiling tool：数据分析工具
⑨ searchable ['sɜːtʃəbl]　adj. 可检索的，可搜索的

management and data governance[①], training on how to navigate the data lake, and an understanding of strong data quality and proper data usage.

3. Benefits of a Data Lake

The data lake offers several benefits, including:

• Developers and data scientists can easily configure[②] a given data model, application, or query on the fly.

• Data lakes are theoretically more accessible. Because there is no inherent[③] structure, any user can technically access the data in the data lake, even though the prevalence[④] of large amounts of unstructured data might inhibit less skilled users.

• The data lake supports users of varying levels of investment; users who want to return to the source to retrieve more information, those who seek to answer entirely new questions with the data and those who simply require a daily report. Access is possible for each of these user types.

• Data lakes are cheap to implement because most technologies used to manage them are open source[⑤] (i.e., Hadoop) and can be installed on low-cost hardware.

• Labor-intensive[⑥] schema development and data cleanup are deferred until after an organization has identified a clear business need for the data.

• The data lake is highly agile. It allows for a variety of different analytics methods to interpret data, including big data analytics, real-time analytics, machine learning and SQL queries.

• The data lake is scalable because of a lack of structure.

4. Data Lake Challenges

The main danger when building a data lake is that bad planning or management can transform the repository into a data swamp[⑦]. A data swamp is a data lake with degraded value, whether due to design mistakes, stale[⑧] data, or uninformed users and lack of regular access. Businesses implementing a data lake should anticipate several important challenges if they wish to avoid being left with a data swamp.

---

① data governance：数据治理

② configure [kən'fɪɡə]　v. 配置；设定

③ inherent [ɪn'hɪərənt]　adj. 固有的，内在的

④ prevalence ['prevələns]　n. 流行；盛行；普遍

⑤ open source：开源

⑥ labor-intensive ['læbərɪnt'ensɪv]　adj. 劳动密集型的

⑦ data swamp：数据沼泽

⑧ stale [steɪl]　adj. 陈旧的，不新鲜的

• Set business priorities: Assuming that any kind of data will eventually provide value and throwing everything into storage is not good practice. Organizations should assess their priorities, then get a general sense for what data is useful to store, and finally anticipate[①] how the business might evolve and what that means for the contents of a potential data lake.

• Designate[②] use cases and end users: Data should be accurate and fit for a purpose, but it should also cater to the people manipulating it. Data inconsistent with[③] the tools and skills available to its consumers serves little purpose.

• Commit to good communication: A data lake cannot be opaque[④] storage. Before implementation, businesses must commit to good communication in order to maintain focus and ensure important stakeholders are aware of how and why to use the data in a data lake. Though data lakes generally benefit from ingestion without modeling, that doesn't mean they shouldn't be documented. Users who know where to look for details regarding the provenance[⑤] and contents of stored data are better prepared to act on that data.

• Establish a robust data ingestion process: Focus on analytics can lead to deemphasizing[⑥] ingestion. Data lakes require fast, accurate ingestion, as getting uncorrupted[⑦] raw data into storage is a primary focus. This step might seem easy when data lakes are concerned, but without a robust data ingestion step, the lake will fail.

---

① anticipate [æn'tɪsɪpeɪt] v. 预期；预计

② designate ['dezɪgneɪt] v. 命名，指定；指派

③ inconsistent with... 与……不符

④ opaque [əʊ'peɪk] adj. 不透明的

⑤ provenance ['prɒvənəns] n. 起源，出处

⑥ deemphasize [diː'efəsaɪz] vt. 降低……的重要性；削弱

⑦ uncorrupted ['ʌnkə'rʌptɪd] adj. 未损坏的

Unit 7

Some Popular Data Science Programming Languages

1. Python

This high-level programming language is one of the most versatile languages as it contains a plethora of libraries that cater to different roles. It is considered easy to use as it is interpreter-based and has high levels of readability. The dynamic language has been around for nearly 30 years now and is used both by small businesses and industry titans like Google, Mozilla, Facebook and Netflix. Indeed also ranks it the third most profitable programming language in the world—yet another reason for it being so popular in the programming community.

Pros of Python:

• Easy to use: Since Python is fully focused on code readability, the language is versatile without being hard to read or understand.

• Open-source: Python is free to download and you can start using it in a matter of few seconds. This is beneficial for everyone but especially for those looking to learn a programming language from scratch but doesn't have the means to buy an expensive course or language package.

• An array of libraries: Whatever you need Python for, the language has a library for it. The most common ones are for machine learning, game development and web development.

Cons of Python:

• Threading problems: According to many users, Python can be tricky when it comes to threading because of the Global Interpreter Lock, which allows only a single thread to execute at a time. The hack is to carry out multiprocessing programmes instead of multiprocessing threads, but this can still be a problem for those looking for threads.

• Not native to mobile: Developers often see Python as weak for mobile computing as it is not native to a mobile environment. It can still be leveraged for the purpose but requires an additional effort that may be beyond the purview of beginners.

2. R

When it comes to exploration of datasets and ad hoc analysis, R scores more points with data scientists. Yet another open-source programming language, R is geared towards statistical computing. It is also a key player in the process of developing numeric analysis and machine learning algorithms. It is often referred to as a "glue language", a reference to its role in connecting datasets, software packages and tools.

Pros of R:

• Reproducible analysis: R is the statistical tool of choice because it produces high-quality data analysis that can be reproduced and scaled. This flexibility allows R to be used on massive datasets and at organisational levels.

• Strong packages: As it was built for statisticians, R has a vast array of packages that can be leveraged to nearly any end to further any statistical technique. Its charting and graphic abilities are also considered to be unmatched.

Cons of R:

• Old design: R is an old language and in that context, it has not had many changes made to its design. This can be a bit problematic for those working with massive datasets, as it has not much kept up with changes in technology or use.

• Lack of inbuilt security: Security was not built into the R language, which means it cannot be embedded into a web browser for secure calculations. It is also difficult to use R like a back-end server for the purpose of building calculations.

3. Java

Java is another object-oriented, general-purpose language. This language tends to be highly versatile and is used in computer embedding, web applications and desktop applications. Java may seem to be disconnected to data science; however, there are many frameworks, including Hadoop, which run on JVM (Java virtual machine) and constitute an integral part of the data stack. Hadoop is a software method for data processing and storage in distributed structures for large data applications. It allows large amounts of data to be processed and possesses the ability to handle virtually limitless tasks at once, thanks to its higher processing power.

Pros of Java:

• Straightforward: Java is one of the lesser complicated languages to learn and is very adaptable to writing, compilation and debugging in the process of development. The code is also reusable and usable in creating standard programmes.

• Distributed computing: In this method, several computers come together on a single network to develop applications simultaneously. Java can be used in such a method, which promotes collaboration over both data and application-related aspects.

• Independent of platforms: Typically, Java code runs on any computer without the need for special software. However, it does need JVM, which allows computers to run both Java programmes and programmes crafted in other languages.

Cons of Java:

• Memory-consuming: Java programmes run on top of JVM, which makes it consume a lot more memory. This could be problematic on systems without much internal memory.

• No support for low-level programming: Although similar to C and C++, Java has fewer low-level facilities in comparison. It is also much slower than these low-level programming languages and cannot support unions and structures.

4. SQL (Structured Query Language)

This domain-specific language is most used for handling data within a relatable database management system. Databases are quite often the backbone of software or an application and are instrumental in determining just how well dependent technologies perform. The more commonly used databases are Oracle, MariaDB, MySQL and PostgreSQL.

Pros of SQL:

• Function-heavy: SQL is well known for being one of the most function-heavy languages but also has a concise syntax. The simpler commands are much easier to understand; however, complex setups and mastering the database's design take a lot more time and effort.

• Speedy for searching and retrieving: Due to the levels of optimisation, SQL databases are said to be the fastest in carrying out data searches over just a single table. With an optimum design, such speeds can easily be achieved even across multiple tables.

Cons of SQL:

• Predefined data model: With SQL databases, data migration becomes an issue. This is because when entering new columns deleting existing ones, every single row in the table gets affected. The way around this is building large-scale migration scripts to adjust existing data for every change.

• Only vertically scalable: Architecturally, SQL databases can only be expanded vertically upon one server. To be able to expand to other servers, more expensive hardware needs to be brought in to the system to be able to cope with massive data dumps and proportionate demands.

5. Scala

Scala has been designed to address many of Java's problems. Again, from web applications to machine learning, this language has many different uses. It is perfect for processing big data.

Pros of Scala:

• Easy to understand: Especially for those with some prior knowledge of Java, Scala's syntax might seem more understandable than any other language. It is also a lot more concise than Java, making it less complicated for beginners looking to write code.

• Scaleable: As the name suggests, Scala is a scalable language. This means it can be easily used to build fault-resistant systems that are concurrent. The fact that it is both object-oriented and functional makes it scalable, as it support higher-order functions, pattern matching and abstractions.

• Concise: Scala is concise and thus provides better support for functions in the back end. However, complexity can be managed by raising the level of abstraction in the existing interfaces.

Cons of Scala:

• Steep learning curve: For developers not familiar with Java, some features like continuations and functional programming might be difficult to process. Though the language spec is much smaller than Java's, the way things are combined is quite unlike Java, which is the source of the relatively steep learning curve.

• Limited developer pool: Scala has fewer developers than Java does, which could be a problem for organisations looking to staff up immediately. It could also be a hindrance for students who are trying to learn Scala and are looking for a mentor or guide. That said, the more the language is explored, the higher the chances of the pool growing in size.

Each of these languages have their indicative purposes, eg., Scala for front-end applications and R for statistical analysis. Thus, the final decision on which programming language to choose depends on the student's field of interest (front-end, statistical analysis, back-end etc.), the uses and benefits of the language in the said field.

New Words

plethora	['pleθərə]	n. 过多，过剩
interpreter-based	[ɪn'tɜːprətə beɪst]	adj. 基于解释器的
readability	[ˌriːdə'bɪlətɪ]	n. 易读性，可读性
rank	[ræŋk]	v. 分等级；排列
profitable	['prɒfɪtəbl]	adj. 有利可图的，有益的；可赚钱的
download	[ˌdaʊn'ləʊd]	v. 下载
thread	[θred]	n. 线程
tricky	['trɪki]	adj. 难办的，难对付的，棘手的
execute	['eksɪkjuːt]	v. 执行，实施
hack	[hæk]	v. 非法侵入(他人计算机系统)
multiprocessing	['mʌltɪˌprəʊsesɪŋ]	n. 多重处理，多处理(技术)
purview	['pɜːvjuː]	n. 眼界，视野，见识
beginner	[bɪ'gɪnə]	n. 新手，初学者
adhoc	[ˌæd'hɒk]	adj. 特别的，特设的；临时的
reproducible	[ˌriːprə'djuːsəbl]	adj. 可再生的，可复写的；能繁殖的
reproduce	[ˌriːprə'djuːs]	v. 复制，重现
unmatched	[ˌʌn'mætʃt]	adj. 无敌的，无比的

problematic	[ˌprɒbləˈmætɪk]	adj. 成问题的，有疑问的
inbuilt	[ˈɪnbɪlt]	adj. 嵌入的，内置的
browser	[ˈbraʊzə]	n. 浏览器
back-end	[bæk end]	adj. 后端的，后台的
object-oriented	[ˈɒbdʒɪkt ˈɔːrɪəntɪd]	adj. 面向对象的
disconnect	[ˌdɪskəˈnekt]	v. 切断；断开
constitute	[ˈkɒnstɪtjuːt]	v. 组成，构成
straightforward	[ˌstreɪtˈfɔːwəd]	adj. 简单明了的；直截了当的
compilation	[ˌkɒmpɪˈleɪʃn]	n. 编译
reusable	[ˌriːˈjuːzəbl]	adj. 可再用的，可重用的
aspect	[ˈæspekt]	n. 方面；样子，外观
independent	[ˌɪndɪˈpendənt]	adj. 独立的
craft	[krɑːft]	v. 精心制作
union	[ˈjuːnɪən]	n. 联合体
instrumental	[ˌɪnstrəˈmentl]	adj. 有帮助的；起作用的
setup	[ˈsetʌp]	n. 设置
script	[skrɪpt]	n. 脚本
adjust	[əˈdʒʌst]	v. 调整，调节
vertically	[ˈvɜːtɪklɪ]	adv. 垂直地
architecturally	[ˌɑːkɪˈtektʃərəli]	adv. 结构上
dump	[dʌmp]	v. 转存
proportionate	[prəˈpɔːʃənət]	adj. 成比例的；相称的
concise	[kənˈsaɪs]	adj. 简明的，简洁的
concurrent	[kənˈkʌrənt]	adj. 同时发生的，并发的
abstraction	[æbˈstrækʃn]	n. 抽象，抽象概念，抽象化
hindrance	[ˈhɪndrəns]	n. 阻碍，障碍
indicative	[ɪnˈdɪkətɪv]	adj. 指示的，表示的
front-end	[frʌnt end]	adj. 前端的

Phrases

data science	数据科学
high-level programming language	高级编程语言，高级程序设计语言
dynamic language	动态语言
in a matter of ...	大约，几个，在……之内
from scratch	从头做起，从零开始
an array of	一排，一群，一批
game development	游戏开发

Global Interpreter Lock	全局解释器锁
carry out	执行，进行，完成
mobile computing	移动计算
be leveraged for	用于
glue language	胶水语言
distributed structure	分布式结构
distributed computing	分布式计算
in comparison	相比之下
concise syntax	简明语法
data migration	数据迁移
cope with	处理，对付，应付
fault-resistant system	防故障系统
higher-order function	高阶函数
pattern matching	模式匹配
learning curve	学习曲线
language spec	语言规范
statistical analysis	统计分析

Abbreviations

JVM (Java Virtual Machine)	Java 虚拟机

Analysis of Difficult Sentences

[1] According to many users, Python can be tricky when it comes to threading because of the Global Interpreter Lock, which allows only a single thread to execute at a time.

本句中，Python 是主语，can be tricky 是谓语。According to many users 和 because of the Global Interpreter Lock 是介词短语，作状语，when it comes to threading 是时间状语从句，它们都修饰谓语。which allows only a single thread to execute at a time 是非限定性定语从句，对 the Global Interpreter Lock 进行补充说明。

[2] Security was not built into the R language, which means it cannot be embedded into a web browser for secure calculations.

本句中，which means it cannot be embedded into a web browser for secure calculations 是非限定性定语从句，对其前面的整个句子进行补充说明。

[3] This is because when entering new columns deleting existing ones, every single row in the table gets affected.

本句中，because when entering new columns deleting existing ones, every single row in the table gets affected 是表语从句。在该从句中，when entering new columns deleting existing ones 是时间状语从句，修饰 gets affected。

[4]   Though the language spec is much smaller than Java's, the way things are combined is quite unlike Java, which is the source of the relatively steep learning curve.

　　本句中，Though the language spec is much smaller than Java's 是让步状语从句，修饰谓语 is quite unlike Java。things are combined 是定语从句，修饰主语 the way。which is the source of the relatively steep learning curve 是非限定性定语从句，对 Java 进行补充说明。

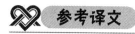 参考译文

一些流行的数据科学编程语言

1. Python

　　这种高级编程语言是用途最广泛的语言之一，它包含大量可满足不同角色的库。Python 基于解释器并且具有高度的可读性，因此易于使用。这种动态语言已经存在近 30 年，小型企业和行业巨头(如 Google、Mozilla、Facebook 和 Netflix)都在使用这种语言。Indeed 还将它列为世界上第三大最赚钱的编程语言——这也是它在编程中受欢迎的另一个原因。

　　Python 的优点如下：

　　• 易于使用：由于 Python 完全专注于代码可读性，因此该语言具有通用性，而且易于阅读或理解。

　　• 开源：Python 可以免费下载，你可以在几秒内开始使用。这对每个人都有好处，对于那些希望零基础学习编程语言但又没有办法购买昂贵的课程或语言包的人来说尤其如此。

　　• 系列库：无论你要用 Python 做什么，该语言都有适合的库。最常见的是机器学习、游戏开发和网络开发。

　　Python 的缺点如下：

　　• 线程问题：许多用户认为，由于全局解释器锁一次只允许执行一个线程，因此 Python 在处理线程时可能会很棘手。黑客攻击是执行多处理程序而不是多处理线程，但这对于那些寻找线程的人来说仍然是一个问题。

　　• 不是移动原生的：开发人员通常认为 Python 对移动计算来说很脆弱，因为它不是原生于移动环境。虽然它仍然可以用于移动计算，但需要额外的努力，这可能超出了初学者的见识。

2. R

　　当涉及数据集的探索和特别分析时，R 在数据科学家中更获认可。R 是另一种开源编程语言，它面向统计计算。它也是开发数值分析和机器学习算法过程中的关键角色。R 通

常被称为"胶水语言"，指的是它在连接数据集、软件包和工具方面的作用。

R 的优点如下：

• 可再现性分析：R 是首选的统计工具，因为它可以生成可再现和缩放的高质量数据分析。这种灵活性使 R 能够在海量数据集和组织级别上使用。

• 强大的包：由于 R 是为统计学家而构建的，所以它拥有大量的软件包，利用这些软件包几乎可以推动任何统计技术的发展。它的图表和图形能力无与伦比。

R 的缺点如下：

• 设计陈旧：R 是一种旧语言，因此，它的设计没有太多变化，因为它跟不上技术或使用的变化。对于处理大量数据集的人来说，这可能有点麻烦。

• 缺乏内置安全性：R 语言没有内置安全性，这意味着它不能嵌入到 Web 浏览器中进行安全计算。也很难像后端服务器一样使用 R 来构建计算。

3．Java

Java 是另一种面向对象的通用语言。这种语言用途广泛，用于计算机嵌入、Web 应用程序和桌面应用程序。Java 似乎与数据科学脱节；但是，有很多框架，包括 Hadoop 都运行在 JVM(Java 虚拟机)上构成了数据栈的组成部分。Hadoop 是一种用于大型数据应用程序的分布式结构中，数据处理和存储的软件方法。由于其拥有更高的处理能力，它能处理大量数据并具有同时处理几乎无限任务的能力。

Java 的优点如下：

• 简单易懂：Java 是一种易于学习的语言，非常适合开发过程中的编码、编译和调试。代码可重用并可用于创建标准程序。

• 分布式计算：在这种方法中，多台计算机聚集在一个网络上同时开发应用程序。Java 可用于这种方法，它促进了数据和应用程序相关方面的协作。

• 独立于平台：通常，Java 代码无需特殊软件即可在任何计算机上运行。但是，它需要 JVM，JVM 允许计算机运行 Java 程序和其他语言编写的程序。

Java 的缺点如下：

• 内存消耗：Java 程序运行在 JVM 之上，这使得它会消耗更多的内存。这在没有太多内部存储器的系统上可能会出现问题。

• 不支持低级编程：虽然与 C 和 C++ 类似，但相比之下 Java 的低级设施较少。它也比这些低级编程语言慢得多，并且不能支持联合体和结构体。

4．SQL(结构化查询语言)

这种特定领域的语言最常用于处理相关数据库管理系统中的数据。数据库通常是软件或应用程序的支柱，有助于确定相关技术的性能。比较常用的数据库有 Oracle、MariaDB、MySQL 和 PostgreSQL。

SQL 的优点如下：

• 功能丰富：众所周知，SQL 是具有最多功能的语言之一，而且语言简洁。SQL 更简单的命令更容易理解；然而，其复杂的设置和掌握数据库的设计需要更多的时间和精力。

• 搜索和检索速度快：由于优化级别，SQL 数据库被认为是对单个表执行数据搜索最快的数据库。通过优化设计，即使跨多个工作表也可以轻松实现这样的速度。

SQL 的缺点如下：

• 预定义数据模型：对于 SQL 数据库，数据迁移是一个问题。这是因为在输入新列删除现有列时，表中的每一行都会受到影响。解决这个问题的方法是构建大规模迁移脚本，以便针对每次更改调整现有数据。

• 只能垂直扩展：在架构上，SQL 数据库只能在一台服务器上垂直扩展。为了能够扩展到其他服务器，需要在系统中引入更昂贵的硬件，以应对海量数据转储和相应的需求。

5. Scala

Scala 旨在解决 Java 的众多问题。而且，从 Web 应用程序到机器学习，这种语言用途多样。它非常适合处理大数据。

Scala 的优点如下：

• 易于理解：特别是对于那些对 Java 有一定了解的人来说，Scala 的语法似乎比其他语言都更容易理解。Scala 也比 Java 简洁得多，对于希望编写代码的初学者来说，它并不复杂。

• 可扩展：顾名思义，Scala 是一种可扩展的语言。这意味着它可以很容易地用于构建并发的抗故障系统。它既是面向对象的语言又是函数语言，这使其具有可扩展性，因为它支持高阶函数、模式匹配和抽象。

• 简洁：可以为后端功能提供更好的支持。但是，它也可以通过提高现有接口中的抽象级别来管理复杂性。

Scala 的缺点如下：

• 陡峭的学习曲线：对于不熟悉 Java 的开发人员，一些特性(如延续和函数式编程)可能难以处理。尽管它的语言规范比 Java 小得多，但组合方式与 Java 大不相同，Java 是相对陡峭的学习曲线的来源。

• 有限的开发人员池：Scala 的开发人员比 Java 少，这对于希望立即增加人员的组织来说可能是问题。对于正在尝试学习 Scala 并寻求指导的学生来说，这也可能是障碍。也就是说，对语言的探索越多，池规模增长的机会就越大。

以上每种语言都有其指示性用途，例如，用于前端应用程序的 Scala 和用于统计分析的 R。因此，最终决定选择哪种编程语言取决于学生感兴趣的领域(前端、统计分析、后端等)，以及该语言在该领域的用途和优势。

Bigdata **Text B**

Hadoop vs. Spark

Hadoop and Spark are two of the most popular processing frameworks for big data architectures. They both provide a rich ecosystem of open source technologies for preparing, processing and managing sets of big data and running analytics applications on them.

1. What Is Hadoop?

First released in 2006, Hadoop was created by software engineers Doug Cutting and Mike Cafarella to process large amounts of data. Hadoop provides a way to efficiently break up large data processing problems across different computers, run computations locally and then combine the results. The architecture makes it easy to build big data applications for clusters containing hundreds or thousands of commodity servers, called nodes.

The main components of Hadoop include the following technologies(see Figure 7-1):

• HDFS. Initially modeled on a file system developed by Google, HDFS manages the process of distributing, storing and accessing data across many separate servers. It can handle both structured and unstructured data, which makes it a suitable choice for building out a data lake.

• YARN. Short for Yet Another Resource Negotiator but typically referred to by its acronym, YARN is Hadoop's cluster resource manager, responsible for executing distributed workloads. It schedules jobs and allocates compute resources such as CPU and memory to applications.

• MapReduce. While its role is reduced by YARN, MapReduce is still the built-in processing engine used to run large-scale batch applications in many Hadoop clusters. It orchestrates the process of splitting large computations into smaller ones that can be spread out across different cluster nodes and then runs the various processing jobs.

• Hadoop Common. This is a set of underlying utilities and libraries used by Hadoop's other components.

Figure 7-1 Components of the Hadoop Ecosystem

2. What Is Spark?

Spark was initially developed by Matei Zaharia in 2009 while he was a graduate student at the University of California, Berkeley. His main innovation with the technology was to improve how data is organized to scale in-memory processing across distributed cluster nodes more efficiently. Like Hadoop, Spark can process vast amounts of data by splitting up workloads on different nodes, but it typically does so much faster. This enables it to handle use cases that Hadoop can't with MapReduce, making Spark more of a general-purpose processing engine.

The following technologies are among Spark's key components:

• Spark Core. This is the underlying execution engine that provides job scheduling and coordinates basic I/O operations, using Spark's basic API.

• Spark SQL. The Spark SQL module enables users to do optimized processing of structured data by directly running SQL queries or using Spark's Dataset API to access the SQL execution engine.

• Spark Streaming and Structured Streaming. These modules add stream processing capabilities. Spark Streaming takes data from different streaming sources, including HDFS, Kafka and Kinesis, and divides it into micro-batches to represent a continuous stream. Structured Streaming is a newer approach built on Spark SQL that's designed to reduce latency and simplify programming.

• MLlib. A built-in machine learning library, MLlib includes a set of machine learning algorithms, plus tools for feature selection and building machine learning pipelines.

3. Architecture

The fundamental architectural difference between Hadoop and Spark relates to how data is organized for processing.

In Hadoop, all the data is split into blocks that are replicated across the disk drives of the various servers in a cluster, with HDFS providing high levels of redundancy and fault tolerance. Hadoop applications can then be run as a single job or a directed acyclic graph (DAG) that contains multiple jobs.

In Hadoop 1.0, a centralized JobTracker service allocates MapReduce tasks across nodes that can run independently of each other, and a local TaskTracker service manages job execution by individual nodes. Starting from Hadoop 2.0, though, JobTracker and TaskTracker are replaced with these components of YARN:

• A ResourceManager daemon that functions as a global job scheduler and resource arbitrator.

• NodeManager, an agent that's installed on each cluster node to monitor resource usage.

• ApplicationMaster, a daemon created for each application that negotiates required resources from ResourceManager and works with NodeManagers to execute processing tasks.

• Resource containers that hold, in an abstract way, the system resources assigned to different nodes and applications.

In Spark, data is accessed from external storage repositories, which could be HDFS, a cloud object store like Amazon Simple Storage Service or various databases and other data sources. While most processing is done in memory, the platform can also "spill" data to disk storage and process it there when data sets are too large to fit into the available memory. Spark can run on clusters managed by YARN, Mesos and Kubernetes or in a standalone mode.

Similar to Hadoop, Spark's architecture has changed significantly from its original design. In early versions, Spark Core organized data into a resilient distributed dataset (RDD), an in-memory data store that is distributed across the various nodes in a cluster. It also created DAGs to help in scheduling jobs for efficient processing.

The RDD API is still supported. But starting with Spark 2.0, which was released in 2016, it was replaced as the recommended programming interface by the Dataset API. Like RDDs, Datasets are distributed collections of data with strong typing features, but they include richer optimizations through Spark SQL to help boost performance.

4. Key Features and Attributes

4.1 Data processing capabilities

Hadoop and Spark are both distributed big data frameworks that can be used to process large volumes of data. Despite the expanded processing workloads enabled by YARN, Hadoop is still oriented mainly to MapReduce, which is well suited for long-running batch jobs that don't have strict service-level agreements.

Spark, on the other hand, typically can run batch workloads as an alternative to MapReduce and also provides higher-level APIs for several other processing use cases. In addition to the SQL, stream processing and machine learning modules, it includes a GraphX API for graph processing and SparkR and PySpark interfaces for R and Python, respectively.

4.2 Performance

Hadoop processing with MapReduce tends to be slow and can be challenging to manage. Spark is often considerably faster for many kinds of batch processing. Proponents claim it can perform up to 100 times faster than an equivalent workload on Hadoop when processing batch jobs in memory.

One big contributor to this is that Spark can do processing without having to write data back to disk storage as an interim step. But even Spark applications written to run on disk can see 10 times faster performance than comparable MapReduce workloads on Hadoop, according to Spark's developers.

But Hadoop may have an advantage when it comes to managing many longer-running workloads on the same cluster simultaneously. Running a lot of Spark applications at the same time can sometimes create memory issues that slow the performance of all the applications.

4.3   Scalability

As a general principle, Hadoop systems can scale to accommodate larger data sets that are sporadically accessed because the data can be stored and processed more cost-effectively on disk drives versus memory. A YARN Federation feature added in Hadoop 3.0, which was released in 2017, enables clusters to support tens of thousands of nodes or more by connecting multiple "subclusters" that have their own resource managers.

The downside is that IT and big data teams may have to invest in more labor for on-premises implementations to provision new nodes and add them to a cluster. Also, with Hadoop, storage is colocated with compute resources on the cluster nodes, which can make it difficult for applications and users outside of the cluster to access the data. But some of these scalability issues can be automatically managed with Hadoop services in the cloud.

One of Spark's main advantages is that storage and compute are separated, which can make it easy for applications and users to access the data from anywhere. Spark includes tools that can help users dynamically scale nodes up and down depending on workload requirements. It's also easier to automatically reallocate nodes at the end of a processing cycle in Spark. A scaling challenge with Spark applications is ensuring that workloads are separated across nodes independent of each other to reduce memory leakage.

5.   Applications and Use Cases

Both Hadoop MapReduce and Spark are often used for batch processing jobs, such as extract, transform and load tasks to move data into a data lake or data warehouse. They both can also handle various big data analytics applications involving recent or historical data, such as customer analytics, predictive modeling, business forecasting, risk management and cyber threat intelligence.

Spark is often a better choice for data streaming and real-time analytics use cases, such as fraud detection, predictive maintenance, stock trading, recommendation engines, targeted advertising and airfare and hotel pricing. It's also typically a better fit for running quick analyses, graph computations and machine learning applications. In addition to including MLlib, Spark is now the recommended back-end platform for Apache Mahout, a machine learning and distributed linear algebra framework that initially was built on top of Hadoop MapReduce.

6.   Deployment and Processing Costs

Organizations can deploy both the Hadoop and Spark frameworks using the free open source versions or commercial cloud services and on-premises offerings. However, the initial deployment costs are just one component of the overall cost of running the big data platforms. IT and data management teams also must include the resources and expertise required to securely provision, maintain and update the underlying infrastructure and big data architecture.

One difference is that a Spark implementation typically will require more memory, which can increase costs when building out a cluster.

The broad Hadoop ecosystem also has a variety of optional supporting technologies to install, configure and maintain, including widely used tools like the HBase database and Hive data warehouse software. Many of them can be used with Spark, too. Commercial versions of the frameworks bundle sets of these components together, which can simplify deployments and may help keep overall costs down.

7. Hadoop or Spark? It isn't Always a Rivalry

Hadoop and Spark aren't mutually exclusive. Sushant Rao, senior director of product marketing at big data platform vendor Cloudera, said that most businesses using Hadoop for data engineering, data preparation, machine learning and other applications are also using Spark as part of those workflows. In addition, both frameworks are commonly combined with other open source components for various tasks.

New Words

rich	[rɪtʃ]	adj. 丰富的
node	[nəʊd]	n. 节点
acronym	['ækrənɪm]	n. 首字母缩略词
micro-batch	['maɪkrəʊ bætʃ]	n.&v. 微批处理
simplify	['sɪmplɪfaɪ]	v. 使简化
block	[blɒk]	n. 块
disk	[dɪsk]	n. 磁盘
daemon	['diːmən]	n. 守护进程
arbitrator	['ɑːbɪtreɪtə]	n. 仲裁
negotiate	[nɪ'gəʊʃɪeɪt]	v. 谈判，协商
container	[kən'teɪnə]	n. 容器
spill	[spɪl]	v. 溢出
standalone	['stændə‚ləʊn]	adj. 单独的，独立的
		n. 脱机
contributor	[kən'trɪbjʊtə]	n. 作出贡献者；促成因素
principle	['prɪnsəpl]	n. 原则，法则
sporadically	[spə'rædɪkli]	adv. 偶发地，零星地
subcluster	[sʌb'klʌstə]	n. 子集群
provision	[prə'vɪʒn]	v. 为……提供所需物品
		n. 提供，供给
dynamically	[daɪ'næmɪkli]	adj. 动态的

mutually	[ˈmjuːtʃʊəli]	adv. 互相地
exclusive	[ɪkˈskluːsɪv]	adj. 排斥的，排外的；独有的

Phrases

software engineer	软件工程师
file system	文件系统
resource manager	资源管理程序，资源管理器
split ... into ...	把……划分为……，把……分成……
general-purpose processing engine	通用处理引擎
multiple job	多作业
abstract way	抽象方式
external storage repository	外部存储库
programming interface	编程接口
long-running batch job	长期运行的批处理作业
service-level agreement	服务级协议
invest in ...	在……上投资，在……投入(时间、精力等)
back-end platform	后端平台
deployment cost	部署成本
management team	管理团队；经营团队

Abbreviations

YARN (Yet Another Resource Negotiator)	另一种资源协调者
CPU (Central Processing Unit)	中央处理器，中央处理单元
I/O (Input/Output)	输入/输出
DAG (Directed Acyclic Graph)	有向非循环图

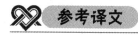 参考译文

Hadoop 与 Spark

Hadoop 和 Spark 是两种最流行的大数据架构处理框架。它们都提供了丰富的开源技术生态系统，用于准备、处理和管理大数据集并在其上运行分析应用程序。

1．什么是 Hadoop？

Hadoop 于 2006 年首次发布，由软件工程师 Doug Cutting 和 Mike Cafarella 创建，用于处理大量数据。Hadoop 提供了能有效解决跨不同计算机的大型数据处理问题、在本地进行计算然后得到组合结果的方法。该架构使为包含数百或数千个商品服务器(被称为节点)的集群构建运行大数据应用程序变得容易。

Hadoop 的主要组件包括以下技术(见图 7-1)：

• HDFS。HDFS 最初以 Google 开发的文件系统为模型，管理跨多个独立服务器分发、存储和访问数据的过程。它可以处理结构化和非结构化数据，成为构建数据湖的合适选择。

• YARN。YARN 是 Yet Another Resource Negotiator(另一种资源协调者)的缩写，但通常以首字母缩写来表示，YARN 是 Hadoop 的集群资源管理器，负责执行分布式工作负载。它调度作业并将计算资源(例如 CPU 和内存)分配给应用程序。

• MapReduce。虽然 YARN 减少了它的作用，但 MapReduce 仍然是内置处理引擎，它在许多 Hadoop 集群中被用来运行大规模批处理应用程序。它协调将大型计算拆分为较小计算的过程，这些较小的计算可以分布在不同的集群节点上，然后运行各种处理作业。

• Hadoop Common。这是一组由 Hadoop 的其他组件使用的底层实用程序和库。

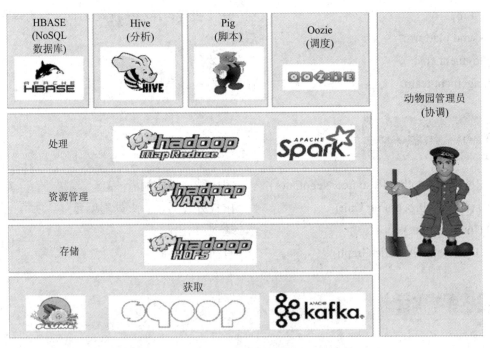

图 7-1　Hadoop 生态系统的组成部分

2．什么是 Spark？

Spark 最初由 Matei Zaharia 于 2009 年开发，当时他还是加州大学伯克利分校的研究生。

他对这项技术的主要创新是改进了数据的组织方式，以便更有效地进行跨分布式集群节点扩展内存处理。与 Hadoop 一样，Spark 可以通过在不同节点上拆分工作负载来处理大量数据，但它的处理速度通常快得多。这使其能够处理 Hadoop 用 MapReduce 所无法处理的用例，使 Spark 更像是通用处理引擎。

以下技术是 Spark 的关键组件：

• Spark Core。这是使用 Spark 的基本 API，它提供底层执行引擎，以实现作业调度和协调基本 I/O 操作。

• Spark SQL。Spark SQL 模块允许用户通过直接运行 SQL 查询来进行结构化数据的优化处理，或者使用 Spark 的 Dataset API 来访问 SQL 执行引擎。

• Spark Streaming 和 Structured Streaming。这些模块添加了流处理功能。Spark Streaming 从不同的流源(包括 HDFS、Kafka 和 Kinesis)获取数据，并将其划分为微批处理以表示连续流。Structured Streaming 是基于 Spark SQL 的较新方法，旨在减少延迟并简化编程。

• MLlib。MLlib 是一个内置的机器学习库，包括一组机器学习算法以及用于特征选择和构建机器学习途径的工具。

3. 构架

Hadoop 和 Spark 之间的基本架构差异与处理数据的组织方式有关。

在 Hadoop 中，所有数据都被分成多个块，这些块在集群中各个服务器的磁盘驱动器之间复制，而 HDFS 提供了高水平的冗余和容错能力。然后，Hadoop 应用程序可以作为单个作业或包含多个作业的有向无环图(DAG)运行。

在 Hadoop1.0 中，集中式 JobTracker 服务在可以相互独立运行的节点之间分配 MapReduce 任务，本地 TaskTracker 服务管理各个节点的作业执行。但是，从 Hadoop2.0 开始，JobTracker 和 TaskTracker 被 YARN 的以下组件替换：

• ResourceManager 守护进程，用作全局作业调度程序和资源仲裁程序。

• NodeManager，一个安装在每个集群节点上的代理，用于监管资源使用情况。

• ApplicationMaster，为每个应用程序创建的守护进程，它与 ResourceManager 协商所需的资源，并与 NodeManagers 一起执行处理任务。

• 以抽象方式保存分配给不同节点和应用程序的系统资源的资源容器。

在 Spark 中，数据是从外部存储库访问的，这些存储库可以是 HDFS、像 Amazon Simple Storage Service 这样的云对象存储或各种数据库和其他数据源。虽然大多数处理是在内存中完成的，但当数据集太大而无法放入可用内存时，平台也可以将数据"溢出"到磁盘存储并在那里进行处理。Spark 可以在 YARN、Mesos 和 Kubernetes 管理的集群上运行，也可以在独立模式下运行。

与 Hadoop 类似，Spark 的架构与其原始设计相比发生了显著变化。在早期版本中，Spark Core 将数据组织成一个弹性分布式数据集(RDD)，这是一种分布在集群中各个节点的内存数据存储。它还创建了 DAG 以便高效处理调度作业。

仍然支持 RDD API。但是从 2016 年发布的 Spark2.0 开始，Dataset API 取代它成为推荐编程接口。与 RDD 一样，数据集是具有强类型功能的分布式数据集合，但它们通过 SparkSQL 包含更丰富的优化以帮助提高性能。

4．主要特点和属性

4.1 数据处理能力

Hadoop 和 Spark 都是分布式大数据框架，可用于处理大量数据。尽管 YARN 支持扩展处理工作负载，Hadoop 仍然主要面向 MapReduce，它非常适合没有严格服务级别协议的长时间运行的批处理作业。

另一方面，Spark 通常可以作为 MapReduce 的替代方案运行批处理工作负载，并且还为其他几个处理用例提供更高级别的 API。除了 SQL、流处理和机器学习模块之外，它还包括用于图形处理的 GraphX API 以及分别用于 R 和 Python 的 SparkR 和 PySpark 接口。

4.2 性能

使用 Hadoop 的 MapReduce 处理速度往往很慢，并且管理起来很有挑战性。对于许多类型的批处理而言，Spark 通常要快得多。Spark 支持者声称，在内存中处理批处理作业时，同等工作负载情况下，Spark 的运行速度比 Hadoop 快 100 倍。

其中一个重要原因是 Spark 可以直接进行处理，而无需作为中间步骤将数据写回磁盘存储。但根据 Spark 的开发人员的说法，即使是在磁盘上运行的 Spark 应用程序，其性能也比 Hadoop 上的类似 MapReduce 工作负载快 10 倍。

但是，在同时管理同一集群上许多运行时间较长的工作负载时，Hadoop 可能具有优势。同时运行大量 Spark 应用程序有时会产生内存问题，从而降低所有应用程序的性能。

4.3 可扩展性

作为一般原则，Hadoop 系统可以扩展来容纳偶尔访问的更大数据集，因为与内存相比，它可以在磁盘驱动器上更经济高效地存储和处理数据。2017 年发布的 Hadoop3.0 中添加了 YARN 联合功能，通过连接多个具有自己的资源管理器的"子集群"，使集群能够支持数万个或更多节点。

不利的一面是，IT 和大数据团队可能必须投入更多的人力用于内部部署实施，以提供新节点并将其添加到集群中。此外，使用 Hadoop，存储于集群节点上的计算资源位于同一位置，这会使集群外的应用程序和用户难以访问数据。但是其中一些可扩展性问题可以通过云中的 Hadoop 服务自动管理。

Spark 的主要优势之一是存储和计算分离，这使得应用程序和用户可以轻松地从任何地方访问数据。Spark 包含的工具可以帮助用户根据工作负载需求动态扩展和缩减节点。在 Spark 中的处理周期结束时自动重新分配节点也更容易。Spark 应用程序的扩展挑战是确保工作负载在彼此独立的节点之间分离，以减少内存泄漏。

5．应用和用例

Hadoop MapReduce 和 Spark 都经常用于批处理作业，例如提取、转换和加载任务以将数据移动到数据湖或数据仓库中。它们还可以处理涉及近期或历史数据的各种大数据分析应用程序，例如客户分析、预测建模、业务预测、风险管理和网络威胁情报。

Spark 通常是数据流和实时分析用例的更好选择，例如欺诈检测、预测性维护、股票交易、推荐引擎、定向广告以及机票和酒店定价。它通常也更适合运行快速分析、图形计算和机器学习应用程序。除了包含 MLlib 之外，Spark 现在还是 ApacheMahout 的推荐后端平台，Apache Mahout 是一种机器学习和分布式线性代数框架，最初构建在 Hadoop MapReduce 之上。

6. 部署和处理成本

企业可以使用免费的开源版本或商业云服务和本地产品来部署 Hadoop 和 Spark 框架。然而，初始部署成本只是运行大数据平台总成本的组成部分。IT 和数据管理团队的工作还必须包括安全地配置、维护和更新底层基础设施和大数据架构所需的资源和专业知识。

区别是 Spark 通常需要更多内存才能实现，这会增加构建集群时的成本。

广泛的 Hadoop 生态系统还有多种可选的支持技术可供安装、配置和维护，包括广泛使用的工具，如 HBase 数据库和 Hive 数据仓库软件。其中许多也可以与 Spark 一起使用。框架的商业版本将这些组件的集合捆绑在一起，这可以简化部署并有助于降低总体成本。

7. Hadoop 还是 Spark? 不总是竞争

Hadoop 和 Spark 并不相互排斥。大数据平台供应商 Cloudera 的产品营销高级总监 Sushant Rao 表示，大多数使用 Hadoop 进行数据工程、数据准备、机器学习和其他应用程序的企业也将 Spark 作为工作流的一部分。此外，这两个框架通常与其他开源组件结合用于各种任务。

Bigdata **Exercises**

〖Ex. 1〗 根据 Text A 回答以下问题。

1. What are the pros of Python?
2. Why do developers often see Python as weak for mobile computing?
3. When does R score more points with data scientists?
4. What are the cons of R?
5. What are the pros of Java?
6. What are the cons of Java?
7. What is this domain-specific language most used for?
8. What is SQL well known?
9. What are the pros of Scala?
10. What are the cons of Scala?

〖Ex. 2〗 根据 Text B 回答以下问题。

1. Who created Hadoop? When was it first released?
2. What do the main components of Hadoop include?
3. What are the technologies among Spark's key components?
4. What does the fundamental architectural difference between Hadoop and Spark relate to?
5. Where is data accessed from in Spark?
6. What can Spark typically do?
7. When may Hadoop have an advantage?
8. What is one of Spark's main advantages?
9. What is Spark often a better choice for?
10. How can organizations deploy both the Hadoop and Spark frameworks?

〖Ex. 3〗 把下列词组翻译成中文。

1. data science   1. _____
2. data migration   2. _____
3. distributed computing   3. _____
4. dynamic language   4. _____
5. programming interface   5. _____
6. mobile computing   6. _____
7. pattern matching   7. _____
8. statistical analysis   8. _____
9. deployment cost   9. _____
10. file system   10. _____

〖Ex. 4〗 把下列单词或词组翻译成英文。

1. 服务级协议   1. _____
2. 资源管理程序，资源管理器   2. _____
3. v. 调整，调节   3. _____
4. n. 浏览器   4. _____
5. n. 编译   5. _____
6. v. 下载   6. _____
7. v. 执行，实施   7. _____
8. adj. 独立的   8. _____
9. adj. 面向对象的   9. _____

10. n. 脚本 _____   10. _____

〖Ex. 5〗 翻译句子。

1. Data science is the field of exploring, manipulating, and analyzing data, and using data to answer questions or make recommendations.
2. The high-level language is a programming language based on English.
3. JAVA is an object-oriented, robust security, portability, dynamic language.
4. User management is an important part of distributed computing environments.
5. Data migration is the process of migrating data from one data environment to another.
6. Statistical analysis is a means of arriving at a fairly reliable forecast.
7. In your Web browser's cache are the most recent Web files that you have downloaded.
8. They had hacked secret data.
9. The bug was caused by an error in the script.
10. This setting controls the size of the file system cache.

Bigdata **Reading Material**

Hadoop Big Data Processing

1. What Is Hadoop?

Hadoop software library[①] is an open-source framework that allows you to efficiently manage and process big data in a distributed[②] computing environment.

2. Main Modules of Hadoop

2.1 Hadoop Distributed File System (HDFS)

Data resides in Hadoop's Distributed File System, which is similar to that of a local file system on a typical computer. HDFS provides better data throughput[③] when compared to traditional file systems.

Furthermore, HDFS provides excellent scalability. You can scale from a single machine to thousands with ease and on commodity hardware.

2.2 Yet Another Resource Negotiator (YARN)

YARN facilitates scheduled tasks, whole managing, and monitoring cluster nodes and other

---

① library ['laɪbrəri] n. 库
② distributed [dɪs'trɪbjuːtɪd] adj. 分布式的
③ throughput ['θruːpʊt] n. 吞吐量；流率

resources.

2.3 MapReduce

The Hadoop MapReduce module helps programs to perform parallel data computation[①]. The Map task of MapReduce converts the input data into key-value pairs[②]. Reduce tasks consume the input, aggregate it, and produce the result.

2.4 Hadoop Common

Hadoop Common uses standard Java libraries across every module.

3. Why Was Hadoop Developed?

The World Wide Web[③] grew exponentially during the last decade, and it now consists of billions of pages. Searching for information online became difficult due to its significant quantity. This data became big data, and it consists of two main problems:

- Difficulty in storing all this data in an efficient and easy-to-retrieve manner.
- Difficulty in processing the stored data.

Developers worked on many open-source projects to return web search results faster and more efficiently by addressing the above problems. Their solution was to distribute data and calculations across a cluster of servers to achieve simultaneous[④] processing.

Eventually, Hadoop came to be a solution to these problems and brought along many other benefits, including the reduction of server deployment cost[⑤].

4. How Is Hadoop Used for Big Data Processing?

Using Hadoop, we utilize the storage and processing capacity of clusters and implement distributed processing for big data. Essentially, Hadoop provides a foundation[⑥] on which you build other applications to process big data.

Applications that collect data in different formats store them in the Hadoop cluster via Hadoop's API, which connects to the NameNode. The NameNode captures the structure of the file directory and the placement of "chunks[⑦]" for each file created. Hadoop replicates these chunks across DataNodes for parallel processing.

MapReduce performs data querying. It maps out all DataNodes and reduces the tasks related to the data in HDFS. The name, "MapReduce" itself describes what it does. Map tasks

---

① parallel data computation：并行数据计算

② key-value pair：关键值对

③ World Wide Web：万维网

④ simultaneous [ˌsɪmlˈteɪnɪəs] adj. 同时的

⑤ deployment cost：部署成本，部署花费

⑥ foundation [faʊnˈdeɪʃn] n. 基础

⑦ chunk [tʃʌŋk] n. 块；大量

run on every node for the supplied input files, while reducers run to link the data and organize the final output.

5. Hadoop's Big Data Tools

Hadoop's ecosystem supports a variety of open-source big data tools. These tools complement Hadoop's core components and enhance its ability to process big data.

The most useful big data processing tools include:

• Apache Hive—a data warehouse for processing large sets of data stored in Hadoop's file system.

• Apache Zookeeper—it automates failovers[1] and reduces the impact of a failed NameNode.

• Apache HBase—an open-source non-relation database for Hadoop.

• Apache Flume—a distributed service for data streaming large amounts of log data.

• Apache Sqoop—a command line[2] tool for migrating data between Hadoop and relational databases.

• Apache Pig—Apache's development platform for developing jobs that run on Hadoop. The software language in use is Pig Latin.

• Apache Oozie—a scheduling system that facilitates the management of Hadoop jobs.

• Apache HCatalog—a storage and table management tool for sorting data from different data processing tools.

6. Advantages of Hadoop

Hadoop is a robust solution for big data processing and is an essential tool for businesses that deal with big data.

The major features and advantages of Hadoop are detailed below.

• Faster storage and processing of vast amounts of data. The amount of data to be stored increases dramatically[3] with the arrival of social media and the Internet of Things (IoT). Storage and processing of these datasets are critical to the businesses that own them.

• Flexibility. Hadoop's flexibility allows you to save unstructured data types such as text, symbols, images, and videos. In traditional relational databases, you will need to process the data before storing it. However, with Hadoop, preprocessing data is not necessary as you can store data as it is and decide how to process it later. In other words, it behaves as a NoSQL database.

• Processing power. Hadoop processes big data through a distributed computing model[4]. Its efficient use of processing power makes it both fast and efficient.

---

① failover [feɪl'əʊvə]　n. 失效备援

② command line：命令行

③ dramatically [drə'mætɪk(ə)li]　adv. 显著地

④ distributed computing model：分布式计算模型

• Reduced cost. Many teams abandoned their projects before the arrival of frameworks like Hadoop, due to the high costs they incurred. Hadoop is an open-source framework, it is free to use, and it uses cheap commodity hardware to store data.

• Scalability. Hadoop allows you to quickly scale your system without much administration, just by merely changing the number of nodes in a cluster.

• Fault tolerance. One of the many advantages of using a distributed data model is its ability to tolerate failures. Hadoop does not depend on hardware to maintain availability. If a device fails[①], the system automatically redirects[②] the task to another device. Fault tolerance is possible because redundant data is maintained by saving multiple copies of data across the cluster. In other words, high availability is maintained at the software layer.

7. The Three Main Use Cases

7.1 Processing big data

We recommend Hadoop for vast amounts of data, usually in the range of petabytes or more. It is better suited for massive amounts of data that require enormous processing power. Hadoop may not be the best option for an organization that processes smaller amounts of data in the range of several hundred gigabytes.

7.2 Storing a diverse[③] set of data

One of the many advantages of using Hadoop is that it is flexible and supports various data types. Irrespective of whether data consists of text, images, or video data, Hadoop can store it efficiently. Organizations can choose how they process data depending on their requirement. Hadoop has the characteristics of a data lake as it provides flexibility over the stored data.

7.3 Parallel data processing

The MapReduce algorithm used in Hadoop orchestrates[④] parallel processing of stored data, meaning that you can execute several tasks simultaneously. However, joint operations are not allowed as it confuses the standard methodology[⑤] in Hadoop. It incorporates parallelism as long as the data is independent of each other.

8. What Are the Challenges of Using Hadoop?

Every application comes with both advantages and challenges. Hadoop also introduces several challenges:

• The MapReduce algorithm isn't always the solution. The MapReduce algorithm does not

---

① fail [feɪl] v. 失败；出故障

② redirect [ˌriːdəˈrekt] vt. 重定向，改变方向

③ diverse [daɪˈvɜːs] adj. 各种的；不同的

④ orchestrate [ˈɔːkɪstreɪt] vt. 把(乐曲)编成管弦乐；精心安排

⑤ methodology [ˌmeθəˈdɒlədʒi] n. 一套方法；方法学，方法论

support all scenarios. It is suitable for simple information requests and issues that be chunked up into independent units, but not for iterative[①] tasks.

MapReduce is inefficient for advanced analytic computing as iterative algorithms require intensive intercommunication[②], and it creates multiple files in the MapReduce phase.

• Hadoop does not provide comprehensive[③] tools for data management, metadata, and data governance. Furthermore, it lacks the tools required for data standardization and determining quality.

Hadoop administration is both an art and a science, requiring low-level knowledge of operating systems, hardware, and Hadoop kernel settings.

• Data security. The Kerberos authentication[④] protocol[⑤] is a significant step towards making Hadoop environments secure. Data security is critical to safeguard big data systems from fragmented data security issues.

Hadoop is highly effective at addressing big data processing when implemented effectively with the steps required to overcome its challenges. It is a versatile[⑥] tool for companies that deal with extensive amounts of data.

---

① iterative ['ɪtərətɪv]   adj. 重复的，反复的，迭代的
② intercommunication [ˌɪntəkəˌmjuːnɪ'keɪʃn]   n. 内部通信，双向通信
③ comprehensive [ˌkɒmprɪ'hensɪv]   adj. 全面的；综合性的
④ authentication [ɔːˌθentɪ'keɪʃn]   n. 身份验证；认证
⑤ protocol ['prəʊtəkɒl]   n. 协议
⑥ versatile ['vɜːsətaɪl]   adj. 多用途的；多功能的

Unit 8

Data Visualization

Data visualization is the practice of translating information into a visual context, such as a map or graph, to make data easier for the human brain to understand and pull insights from. The main goal of data visualization is to make it easier to identify patterns, trends and outliers in large data sets. The term is often used interchangeably with others, including information graphics, information visualization and statistical graphics.

Data visualization is one of the steps of the data science process, which states that after data has been collected, processed and modeled, it must be visualized for conclusions to be made. Data visualization is also an element of the broader data presentation architecture (DPA) discipline, which aims to identify, locate, manipulate, format and deliver data in the most efficient way possible.

1.　Why Is Data Visualization Important?

Data visualization is important for almost every career. It can be used by teachers to display student test results, by computer scientists exploring advancements in artificial intelligence (AI) or by executives looking to share information with stakeholders. It also plays an important role in big data projects. As businesses accumulated massive collections of data during the early years of the big data trend, they needed a way to quickly and easily get an overview of their data. Visualization tools were a natural fit.

Visualization is central to advanced analytics for similar reasons. When a data scientist is writing advanced predictive analytics or machine learning (ML) algorithms, it becomes important to visualize the outputs to monitor results and ensure that models are performing as intended. This is because visualizations of complex algorithms are generally easier to interpret than numerical outputs.

Data visualization provides a quick and effective way to communicate information in a universal manner using visual information. The practice can also help businesses identify which factors affect customer behavior; pinpoint areas that need to be improved or need more attention; make data more memorable for stakeholders; understand when and where to place specific products and predict sales volumes.

Other benefits of data visualization include the following:

- The ability to absorb information quickly, improve insights and make faster decisions;
- An increased understanding of the next steps that must be taken to improve the organization;
- An improved ability to maintain the audience's interest with information they can understand;
- An easy distribution of information that increases the opportunity to share insights with everyone involved;
- Eliminating the need for data scientists since data is more accessible and understandable;
- An increased ability to act on findings quickly and, therefore, achieve success with greater speed and less mistakes.

2. Disadvantages of Data Visualization

The increased popularity of big data and data analysis projects have made visualization more important than ever. Companies are increasingly using machine learning to gather massive amounts of data that can be difficult and slow to sort through, comprehend and explain. Visualization offers a means to speed this up and present information to business owners and stakeholders in ways they can understand.

Big data visualization often goes beyond the typical techniques used in normal visualization, such as pie charts, histograms and corporate graphs. It uses more complex representations, such as heat maps instead. Big data visualization requires powerful computer systems to collect raw data, process it and turn it into graphical representations that humans can use to quickly draw insights.

While big data visualization can be beneficial, it can pose several disadvantages to organizations. They are as follows:

- To get the most out of big data visualization tools, a visualization specialist must be hired. This specialist must be able to identify the best data sets and visualization styles to guarantee organizations are optimizing the use of their data.
- Big data visualization projects often require involvement from IT as well as management, since the visualization of big data requires powerful computer hardware, efficient storage systems and even a move to the cloud.
- The insights provided by big data visualization will only be as accurate as the information being visualized. Therefore, it is essential to have people and processes in place to govern and control the quality of corporate data, metadata and data sources.

3. Examples of Data Visualization

In the early days of visualization, the most common visualization technique was using a Microsoft Excel spreadsheet to transform the information into a table, bar graph or pie chart. While these visualization methods are still commonly used, more intricate techniques are now available, including infographics, bubble clouds, bullet graphs, heat maps and time series charts.

Some other popular techniques are as follows:

• Line charts. This is one of the most basic and common techniques used. Line charts display how variables can change over time.

• Area charts. This visualization method is a variation of a line chart; it displays multiple values in a time series—or a sequence of data collected at consecutive, equally spaced points in time.

• Scatter plots. This technique displays the relationship between two variables. A scatter plot takes the form of an x- and y-axis with dots to represent data points (see Figure 8-1).

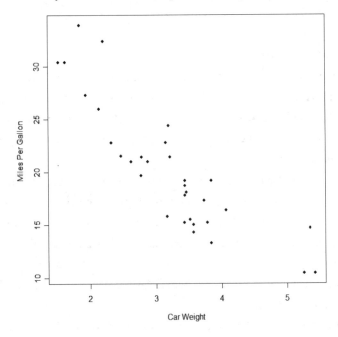

Figure 8-1   Scatter Plots

• Treemaps. This method shows hierarchical data in a nested format. The size of the rectangles used for each category is proportional to its percentage of the whole. Treemaps are best used when multiple categories are present, and the goal is to compare different parts of a whole.

• Population pyramids. This technique uses a stacked bar graph to display the complex narrative of a population. It is best used when trying to display the distribution of a population (see Figure 8-2).

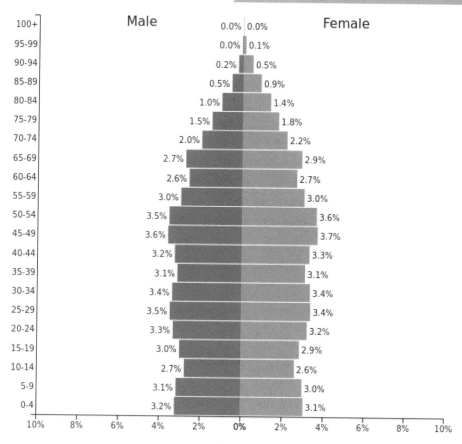

Figure 8-2　Population Pyramids

4. Common Data Visualization Use Cases

Common use cases for data visualization include the following:

• Sales and marketing. Research from the media agency Magna predicts that half of all global advertising dollars will be spent online in three years. As a result, marketing teams must pay close attention to their sources of web traffic and how their web properties generate revenue. Data visualization makes it easy to see traffic trends over time as a result of marketing efforts.

• Healthcare. Healthcare professionals frequently use choropleth maps to visualize important health data. A choropleth map displays divided geographical areas that are assigned a certain color in relation to a numeric variable. Choropleth maps allow professionals to see how a variable, such as the mortality rate of heart disease, changes across specific territories.

• Science. Scientific visualization, sometimes referred to in shorthand as SciVis, allows scientists and researchers to gain greater insight from their experimental data than ever before.

• Finance. Finance professionals must track the performance of their investment decisions when choosing to buy or sell an asset. Candlestick charts are used as trading tools and help finance professionals analyze price movements over a period of time. They display important information, such as securities, derivatives, currencies, stocks, bonds and commodities. By analyzing how the price has changed over time, data analysts and finance professionals can detect trends.

• Logistics. Shipping companies can use visualization tools to determine the best global shipping routes.

• Data scientists and researchers. Visualizations built by data scientists are typically for the scientist's own use, or for presenting the information to a select audience. The visual representations are built using visualization libraries of the chosen programming languages and tools. Data scientists and researchers frequently use open source programming languages—such as Python—or proprietary tools designed for complex data analysis. The data visualization performed by these data scientists and researchers helps them understand data sets and identify patterns and trends that would have otherwise gone unnoticed.

New Words

practice	['præktɪs]	n. 实践；实际行动；通常的做法
		v. 练习
visual	['vɪʒʊəl]	adj. 视觉的
graph	[græf]	n. 图表
interchangeably	[ɪntə'tʃeɪndʒəbli]	adv. 可交换地，可交替地
discipline	['dɪsəplɪn]	n. 学科
career	[kə'rɪə]	n. 职业；事业；生涯
advancement	[əd'vɑːnsmənt]	n. 发展，推动；提升
stakeholder	['steɪkhəʊldə]	n. 股东；利益相关者
accumulate	[ə'kjuːmjʊleɪt]	v. 积累，积聚；堆积
universal	[juːnɪ'vɜːsl]	adj. 通用的，普遍(存在)的
		n. 普遍原则，通用原理
memorable	['memərəbl]	adj. 显著的；难忘的；重大的
absorb	[əb'sɔːb]	v. 吸收；理解，掌握；合并
opportunity	[ˌɒpə'tjuːnəti]	n. 机会，时机
eliminate	[ɪ'lɪmɪneɪt]	v. 排除，清除
histogram	['hɪstəgræm]	n. 直方图，柱状图
beneficial	[ˌbenɪ'fɪʃl]	adj. 有益的，有帮助的
specialist	['speʃəlɪst]	n. 专家
style	[staɪl]	n. 方式；样式
guarantee	[ˌgærən'tiː]	v. 保证，担保
involvement	[ɪn'vɒlvmənt]	n. 参与
intricate	['ɪntrɪkət]	adj. 错综复杂的；难理解的
infographic	['ɪnfəʊ'græfɪk]	n. 信息图

variation	[ˌveərɪˈeɪʃn]	n. 变体；变化，变动
consecutive	[kənˈsekjətɪv]	adj. 连续的，不间断的
treemap	[triːmæp]	n. 树状图
rectangle	[ˈrektæŋgl]	n. 长方形，矩形
proportional	[prəˈpɔːʃənl]	adj. 成比例的
percentage	[pəˈsentɪdʒ]	n. 百分比，百分率
pyramid	[ˈpɪrəmɪd]	n. 金字塔
narrative	[ˈnærətɪv]	n. 描述，叙述
distribution	[ˌdɪstrɪˈbjuːʃn]	n. 分配，分布
property	[ˈprɒpəti]	n. 特性；属性
display	[dɪˈspleɪ]	v. 显示
		n. 显示器
track	[træk]	v. 跟踪
asset	[ˈæset]	n. 资产
security	[sɪˈkjʊərəti]	n. 证券
derivative	[dɪˈrɪvətɪv]	n. 衍生品
stock	[stɒk]	n. 股票；库存
bond	[bɒnd]	n. 债券
commodity	[kəˈmɒdəti]	n. 商品
logistics	[ləˈdʒɪstɪks]	n. 物流；后勤
route	[ruːt]	n. 常规路线；航线；渠道，途径
unnoticed	[ˌʌnˈnəʊtɪst]	adj. 未被觉察到的，未被注意到的；被忽视的

Phrases

translate ... into ...	把……转换为……
pull ... from ...	从……拉出来……
information visualization	信息可视化
statistical graphic	统计图形
customer behavior	客户行为
pie chart	饼图
heat map	热图
computer system	计算机系统
raw data	原始数据
storage system	存储系统
bubble cloud	气泡云
bullet graph	子弹图
time series chart	时间序列图

line chart	折线图
area chart	面积图
scatter plot	散点图
nested format	嵌套格式
web traffic	网络流量
choropleth map	地区分布图
numeric variable	数字变量

Abbreviations

DPA (Data Presentation Architecture)　　　数据呈现架构

 ## Analysis of Difficult Sentences

[1]　Data visualization is one of the steps of the data science process, which states that after data has been collected, processed and modeled, it must be visualized for conclusions to be made.

　　　本句中，which states that after data has been collected, processed and modeled, it must be visualized for conclusions to be made 是非限定性定语从句，对 one of the steps of the data science process 进行补充说明。

[2]　Data visualization is also an element of the broader data presentation architecture (DPA) discipline, which aims to identify, locate, manipulate, format and deliver data in the most efficient way possible.

　　　本句中，which aims to identify, locate, manipulate, format and deliver data in the most efficient way possible 是非限定性定语从句，对 element of the broader data presentation architecture (DPA) discipline 进行补充说明。

[3]　When a data scientist is writing advanced predictive analytics or machine learning (ML) algorithms, it becomes important to visualize the outputs to monitor results and ensure that models are performing as intended.

　　　本句中，When a data scientist is writing advanced predictive analytics or machine learning (ML) algorithms 是时间状语从句，修饰谓语 becomes important。it 是形式主语，真正的主语是动词不定式短语 to visualize the outputs to monitor results and ensure that models are performing as intended。在该短语中，to monitor results and ensure that models are performing as intended 作目的状语。

[4]　Therefore, it is essential to have people and processes in place to govern and control the quality of corporate data, metadata and data sources.

　　　本句中，it 是形式主语，真正的主语是动词不定式短语 to have people and processes

in place to govern and control the quality of corporate data, metadata and data sources。

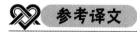 参考译文

数据可视化

数据可视化是将信息转换为可视内容(例如示意图或图表)的行为，使人脑更容易理解数据并从中领悟。数据可视化的主要目标是更容易识别大型数据集中的模式、趋势和异常值。该术语经常与其他术语互换使用，包括信息图形、信息可视化和统计图形。

数据可视化是数据科学过程的步骤之一，它规定在收集、处理和建模数据后，必须将其可视化才能得出结论。数据可视化也是更广泛的数据呈现架构(DPA)学科的元素，该学科旨在以最有效的方式识别、定位、操作、格式化和交付数据。

1. 为什么数据可视化很重要?

数据可视化几乎对所有职业都很重要。教师可以用它来展示学生的考试结果，计算机科学家可以用它来探索人工智能(AI)的进步，管理人员也可以用它与股东分享信息。数据可视化还在大数据项目中发挥着重要作用。由于企业在大数据趋势的早期积累了大量数据，他们需要一种快速、轻松地获得数据概览的方法。可视化工具非常适合。

同样，可视化是高级分析的核心。当数据科学家在编写高级预测分析或机器学习 (ML)算法时，输出可视化监控结果并确保模型按预期执行就变得很重要了。这是因为复杂算法的可视化通常比数字输出更容易解释。

数据可视化提供了一种使用视觉的通用方式交流信息的快速有效的方法。这种方法还可以帮助企业确定哪些因素会影响客户行为;指出需要改进或更多关注的领域;让股东更容易记住数据;了解何时何地放置特定产品并预测销量。

数据可视化还有如下所示的其他好处:
- 能够快速吸收信息、提高洞察力并更快做出决策;
- 进一步了解为改进企业必须采取的后续步骤;
- 提高了能让听众理解信息，从而保持兴趣的能力;
- 轻松发送信息，增加与所有相关人员分享见解的机会;
- 无需数据科学家，因为数据更易访问和理解;
- 提高对结果迅速采取行动的能力，从而以更快速和更少犯错得成功。

2. 数据可视化的缺点

大数据和数据分析项目的日益普及使得可视化比以往任何时候都更加重要。公司越来越多地使用机器学习来收集大量数据，这些数据整理、理解和解释起来又难又慢。可视化提供了一种可行方式来加速这一过程，并以企业主和股东可以理解的方式呈现信息。

大数据可视化通常超出了正常可视化中使用的典型技巧，例如饼图、直方图和公司图

表。它使用更复杂的表示，例如热图。大数据可视化需要强大的计算机系统来收集原始数据、对其进行处理并将其转换为人类可以用来快速得出见解的图形表示。

虽然大数据可视化是有益的，但也可能会给组织带来一些不利因素。详情如下：

• 要充分利用大数据可视化工具，必须聘请可视化专家。该专家必须能够确定最佳数据集和可视化样式，以确保企业优化使用数据。

• 大数据可视化项目通常需要 IT 和管理人员参与，因为大数据可视化需要强大的计算机硬件、高效的存储系统，甚至需要迁移到云端。

• 大数据可视化提供的见解与可视化信息一样准确。因此，必须要有人员和流程来管理和控制企业数据、元数据和数据源的质量。

3．数据可视化示例

早期，最常见的可视化技术是使用 Microsoft Excel 电子表格将信息转换为表格、条形图或饼图。虽然这些可视化方法现在仍然普遍使用，但更复杂的技术，包括信息图、气泡云、子弹图、热图和时间序列图也可以使用。

其他一些流行的可视化技术如下：

• 折线图。这是最基本和最常用的技术之一。折线图显示变量如何随时间变化。

• 面积图。这种可视化方法是折线图的变体；它显示时间序列中的多个值——或在连续等隔时间点收集的数据序列。

• 散点图。该技术显示两个变量之间的关系。散点图采用 x 轴和 y 轴的形式，用点表示数据点(图 8-1)。

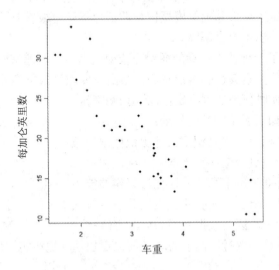

图 8-1　散点图

• 树状图。该方法以嵌套格式显示分层数据。每个类别所用矩形的大小与其在整体中的百分比成正比。当存在多个类别时，最好使用树状图，目的是比较整体的不同部分。

• 人口金字塔。该技术使用堆叠条形图来描述复杂的社会人口描述。在尝试显示人口分布时使用效果最佳(见图 8-2)。

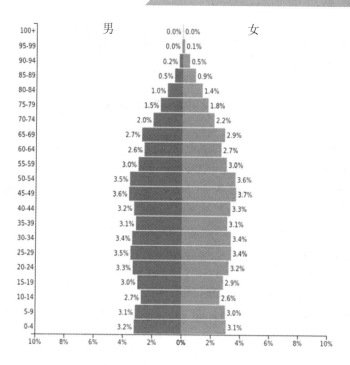

图 8-2　人口金字塔

4．常见数据可视化用例

数据可视化的常见用例包括：

• 销售和营销。根据媒体机构 Magna 的研究预测，三年以后，全球一半的广告费用将用于在线广告。因此，营销团队必须密切关注网络流量来源以及网络资产如何产生收入。通过数据可视化，可以很容易地看到随着时间变化的流量趋势，这是营销工作的结果。

• 医疗保健。医疗保健专业人员经常使用地区分布图来可视化重要的健康数据。地区分布图显示划分的地理区域，这些区域被分配了与数字变量相关的特定颜色。地区分布图让专业人士可以看到变量(例如心脏病死亡率)如何在特定地区发生变化。

• 科学。科学可视化，有时简称为 SciVis，它使科学家和研究人员能够从实验数据中获得比以往任何时候都更深刻的见解。

• 金融。金融专业人士在选择购买或出售资产时必须跟踪其投资决策的表现。K 线图作为交易工具，帮助金融专业人士分析一段时间内的价格走势。它们可以显示重要信息，例如证券、衍生品、货币、股票、债券和商品。通过分析价格如何随时间变化，数据分析师和金融专业人士可以发现趋势。

• 物流。航运公司可以使用可视化工具来确定最佳的全球航运路线。

• 数据科学家和研究人员。数据科学家构建的可视化工具应用通常供科学家自己使用，或用于向选定的受众展示信息。可视化展示是使用所选编程语言和工具的可视化库构建的。数据科学家和研究人员经常使用开源编程语言，例如 Python，或专为复杂数据分析设计的专有工具。这些数据科学家和研究人员执行的数据可视化可以帮助他们理解数据集并识别本来可能会被忽视的模式和趋势。

Bigdata　**Text B**

10 Best Data Visualization Tools

In this era of data, it is very important to understand the data to obtain some actionable insights. And data visualization is a very important part of understanding the hidden patterns and layers in the data! A beautiful and descriptive bar chart or a boring spreadsheet telling the same information? What sounds more interesting to you?

It's, of course, the bar chart as humans are visual creatures and data visualization charts like bar charts, scatter plots, line charts, geographical maps, etc. are extremely important. They tell you information just by looking at them whereas normally you would have to read spreadsheets or text reports to understand the data. Data visualization tools are very popular as they allow analysts and statisticians to create visual data models easily according to their specifications by conveniently providing an interface, database connections, and machine learning tools all in one place!

1. Tableau

Tableau is a data visualization tool that can be used by data analysts, scientists, statisticians, etc. to visualize the data and get a clear opinion based on the data analysis. Tableau is very famous as it can take in data and produce the required data visualization output in a very short time. And it can do this while providing the highest level of security with a guarantee to handle security issues as soon as they arise or are found by users (see Figure 8-3).

Figure 8-3　Tableau Interface

Tableau also allows its users to prepare, clean and format their data and then create data visualizations to obtain actionable insights that can be shared with other users. Tableau is available for the individual data analyst, business teams and organizations. It provides a 14-day free trial followed by the paid version.

2. Looker

Looker is a data visualization tool that can go in-depth in the data and analyze it to obtain useful insights. It provides real-time dashboards of the data for more in-depth analysis so that businesses can make instant decisions based on the data visualizations obtained. Looker also provides connections with Redshift, Snowflake, BigQuery, as well as more than 50 SQL supported dialects so you can connect to multiple databases without any issues.

Looker can be shared with anyone using any particular tool. Also, you can export these files in any format immediately. It also provides customer support wherein you can ask any question and it shall be answered.

3. Zoho Analytics

Zoho Analytics is a business intelligence and data analytics software that can help you create wonderful looking data visualizations based on your data in a few minutes. You can obtain data from multiple sources and mesh it together to create multidimensional data visualizations that allow you to view your business data across departments. In case you have any questions, you can use Zia which is a smart assistant created using artificial intelligence, machine learning, and natural language processing.

Zoho Analytics allows you to share or publish your reports with your colleagues and add comments or engage in conversations as required. You can export Zoho Analytics files in any format such as spreadsheet, MS Word, Excel, PPT, PDF, etc.

4. Sisense

Sisense is a business intelligence-based data visualization system and it provides various tools that allow data analysts to simplify complex data and obtain insights for their organization and outsiders. Sisense believes that eventually, every company will be a data-driven company and every product will be related to data in some way. Therefore it tries its best to provide various data analytics tools to business teams and data analytics so that they can help make their companies the data-driven companies of the future.

It is very easy to set up and learn Sisense. It can be easily installed within a minute and data analysts can get their work done and obtain results instantly. Sisense also allows its users to export their files in multiple formats such as PPT, Excel, MS Word, PDF, etc. Sisense also provides full-time customer support services whenever users face any issues.

5. IBM Cognos Analytics

IBM Cognos Analytics is an artificial intelligence-based business intelligence platform that supports data analytics among other things. You can visualize as well as analyze your data and share actionable insights with anyone in your organization. Even if you have limited or no knowledge about data analytics, you can use IBM Cognos Analytics easily as it interprets the data for you and presents you with actionable insights in plain language.

You can also share your data with multiple users if you want on the cloud and share visuals over email or Slack. You can also import data from various sources like spreadsheets, cloud, CSV files, or on-premises databases and combine related data sources into a single data module.

6. Qlik Sense

Qlik Sense is a data visualization platform that helps companies to become data-driven enterprises. By providing an associative data analytics engine, sophisticated artificial intelligence system, and scalable multi-cloud architecture, you can deploy any combination of SaaS, on-premises or a private cloud.

You can easily combine, load, visualize, and explore your data on Qlik Sense, no matter its size. All the data charts, tables, and other visualizations are interactive and instantly update themselves according to the current data context. The Qlik Sense AI can even provide you with data insights and help you create analytics using just drag and drop.

7. Domo

Domo is a business intelligence model that contains multiple data visualization tools. These tools provide a consolidated platform where you can perform data analysis and then create interactive data visualizations that allow other people to easily understand your data conclusions. You can combine cards, text, and images in the Domo dashboard so that you can guide other people through the data while telling a data story as they go.

In case of any doubts, you can use their pre-built dashboards to obtain quick insights from the data. Domo has a free trial option so you can use it to get a sense of this platform before committing to it fully. In case of any customer service inquiries, Domo is always available from 7 AM to 6 PM from Monday to Friday and you can try it for free followed by the paid version.

8. Microsoft Power BI

Microsoft Power BI is a data visualization platform focused on creating a data-driven business intelligence culture in all companies today. To fulfill this, it offers self-service analytics tools that can be used to analyze, aggregate, and share the data in a meaningful fashion.

Microsoft Power BI offers hundreds of data visualizations to its customers along with built-in artificial intelligence capabilities and Excel integration facilities. It also provides you with multiple support systems such as FAQs, forums and also live chat support with the staff.

9. Klipfolio

Klipfolio is one of the best data visualization tools. You can access your data from hundreds of different data sources like spreadsheets, databases, files, and web services applications by using connectors. Klipfolio also allows you to create custom drag-and-drop data visualizations wherein you can choose from different options like charts, graphs, scatter plots, etc.

Klipfolio also has tools you can use to execute complex formulas that can solve challenging data problems. In the case of customer inquiries, you can get help from the community forum or the knowledge forum.

10. SAP Analytics Cloud

SAP Analytics Cloud uses business intelligence and data analytics capabilities to help you evaluate your data and create visualizations in order to predict business outcomes. It also provides you with the latest modeling tools that help you by alerting you of possible errors in the data and categorizing different data measures and dimensions. SAP Analytics Cloud also suggests Smart Transformations to the data that lead to enhanced visualizations.

In case you have any doubts or business questions related to data visualization, SAP Analytics Cloud provides you with complete customer satisfaction by handling your queries using conversational artificial intelligence and natural language technology.

New Words

hidden	['hɪdn]	adj. 隐藏的
descriptive	[dɪ'skrɪptɪv]	adj. 描述的
boring	['bɔːrɪŋ]	adj. 令人厌烦的，无聊的
creature	['kriːtʃə]	n. 生物，动物
specification	[ˌspesɪfɪ'keɪʃn]	n. 规范，规格；详述；说明书
opinion	[ə'pɪnjən]	n. 意见
famous	['feɪməs]	adj. 著名的，有名的
arise	[ə'raɪz]	v. 出现，产生
individual	[ˌɪndɪ'vɪdʒuəl]	adj. 单独的；个人的
trial	['traɪəl]	v. 测试，试用
version	['vɜːʃn]	n. 版本
dialect	['daɪəlekt]	n. 方言，土话
export	[ɪk'spɔːt]	v. (计算机)导出
mesh	[meʃ]	v. 组成网络；紧密配合
department	[dɪ'pɑːtmənt]	n. 部门
comment	['kɒment]	v. 发表意见

		n. 评论；解释
outsider	[ˌaʊtˈsaɪdə]	n. 局外人，圈外人
install	[ɪnˈstɔːl]	v. 安装
interpret	[ɪnˈtɜːprɪt]	v. 解释，说明
associative	[əˈsəʊʃɪətɪv]	adj. 联合的，结合的
combination	[ˌkɒmbɪˈneɪʃn]	n. 结合(体)，联合(体)
consolidate	[kənˈsɒlɪdeɪt]	v. 统一，合并；加强，巩固
doubt	[daʊt]	n. 疑问
		v. 怀疑；不相信
culture	[ˈkʌltʃə]	n. 文化；文明
forum	[ˈfɔːrəm]	n. 论坛
connector	[kəˈnektə]	n. 连接程序，连接器
outcome	[ˈaʊtkʌm]	n. 结果
suggest	[səˈdʒest]	v. 建议，提议
satisfaction	[ˌsætɪsˈfækʃn]	n. 满足，称心；妥善处理
conversational	[ˌkɒnvəˈseɪʃənl]	adj. 会话的，谈话的；[计]对话的，双向的

Phrases

visualization output	可视化输出
business team	业务团队
paid version	付费版本
customer support	客户支持
smart assistant	智能助手
data-driven company	数据驱动公司
multiple format	多种格式
full-time customer support service	全职的客户支持服务
multiple user	多用户
data analytics engine	数据分析引擎
business intelligence culture	商业智能文化
self-service analytics tool	自助分析工具
community forum	社区论坛
modeling tool	建模工具

Abbreviations

FAQ (Frequently Asked Questions)	常见问题解答

10 个最佳数据可视化工具

在这个数据时代，理解数据获得可操作的见解非常重要。而数据可视化是理解数据中隐藏的模式和层次的非常重要的部分！一个漂亮的描述性条形图还是一个讲述相同信息的乏味电子表格？你觉得哪个更有趣？

当然是条形图，因为人类是视觉生物，条形图、散点图、折线图、地理地图等数据可视化图表非常重要。只要看图就能理解那些必须通过阅读电子表格或文本报告才能理解数据。数据可视化工具非常受欢迎，因为它们允许分析师和统计学家通过方便地提供一个界面、数据库连接和机器学习工具，从而根据他们的规范轻松创建可视化数据模型！

1. Tableau

Tableau 是一种数据可视化工具，可供数据分析师、科学家、统计学家等使用，将数据可视化并基于数据分析得出清晰的意见。Tableau 非常有名，因为它可以在很短的时间内接收数据并生成所需的数据可视化输出。在做到这一点的同时，它还可以提供最高级别的安全性，并保证出现安全问题或用户发现安全问题时立即处理(见图 8-3)。

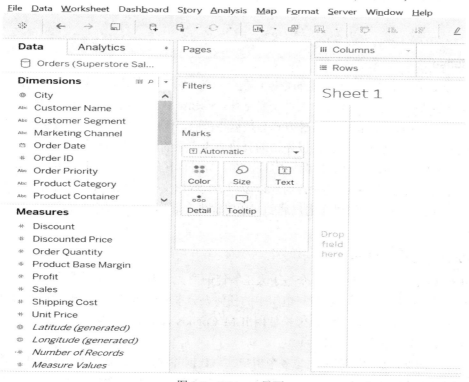

图 8-3 Tableau 界面

Tableau 还允许用户准备、清理和格式化数据，然后创建数据可视化以获得可与其他用户共享的可操作见解。Tableau 可供个人数据分析师、业务团队和组织使用。在付费使用前，Tableau 提供 14 天免费试用。

2. Looker

Looker 是一个可以深入数据并对其进行分析以获得有用的见解的数据可视化工具。Looker 提供了数据实时仪表板以便进行更深入的分析，从而使企业能够根据获得的数据可视化做出即时决策。Looker 还提供与 Redshift、Snowflake、BigQuery 以及 50 多种支持 SQL 的方言的连接，因此你可以无障碍地连接到多个数据库。

Looker 可以与使用任何特定工具的人共享。此外，你可以立即以任何格式导出文件。Looker 还提供客户支持，你可以在其中提出任何问题并得到答复。

3. Zoho Analytics

Zoho Analytics 是一款帮助你在几分钟内根据数据创建精美的数据可视化的商业智能和数据分析软件。可以从多个来源获取数据并将其组合在一起以创建多维数据可视化，允许跨部门查看业务数据。如果有任何疑问，可以使用 Zia，它是使用人工智能、机器学习和自然语言处理创建的智能助手。

Zoho Analytics 允许你与同事共享或发布报告，并根据需要添加评论或参与对话。你可以以任何格式导出 Zoho Analytics 文件，例如电子表格、MS Word、Excel、PPT、PDF 等格式。

4. Sisense

Sisense 是一个基于商业智能的数据可视化系统，它提供了各种工具，允许数据分析师简化复杂的数据并为组织和外部人员获取见解。Sisense 认为，最终每家公司都会成为数据驱动的公司，每款产品都会以某种方式与数据相关联。因此，它竭尽全力为业务团队和数据分析提供各种数据分析工具，以便帮助公司成为未来的数据驱动型公司。

设置和学习 Sisense 非常容易。它可以在一分钟内轻松安装，数据分析师可以立即完成工作并获得结果。Sisense 还允许用户以多种格式导出文件，例如 PPT、Excel、MS Word、PDF 等格式。Sisense 还可以在用户遇到问题时提供全职客户支持服务。

5. IBM Cognos Analytics

IBM Cognos Analytics 是一个基于人工智能的商业智能平台，支持数据分析。你能可视化数据并对其进行分析，也能与组织中的任何人分享可操作的见解。即使你对数据分析知之甚少甚至一无所知，也可以轻松使用 IBM Cognos Analytics，它会解释数据并以通俗易懂的语言提供可操作的见解。

如果你愿意，还可以在云上与多个用户共享数据并通过电子邮件或 Slack 共享视觉效果。你还可以从电子表格、云、CSV 文件或本地数据库等各种来源导入数据，并将相关数据源合并到一个数据模块中。

6. Qlik Sense

Qlik Sense 是一个可帮助企业成为数据驱动型企业的数据可视化平台。通过提供关联数据分析引擎、复杂的人工智能系统和可扩展的多云架构，可以部署 SaaS、本地部署或私有云的任意组合。

无论数据大小，你都可以在 Qlik Sense 上轻松组合、加载、可视化和探索。所有数据图表、表格和其他可视化数据都是交互式的，并根据当前数据状况即时更新。Qlik Sense AI 甚至可以提供数据见解并通过拖放创建分析。

7. Domo

Domo 是一种包含多种数据可视化工具的商业智能模型。这些工具提供了统一的平台，你可以在其中执行数据分析，然后创建交互式数据可视化，让其他人可以轻松理解你的数据结论。你可以在 Domo 仪表板中组合卡片、文本和图像，这样就可以指导其他人浏览数据，同时讲述数据故事。

如有任何疑问，可以使用预先构建的仪表板从数据中快速获取见解。Domo 有一个免费试用选项，可以在投入使用之前试用它来了解这个平台。Domo 在周一至周五的上午 7 点至下午 6 点都提供客户服务查询服务，可以先免费试用，然后再使用付费版本。

8. Microsoft Power BI

Microsoft Power BI 是一个专注于在所有公司中创建数据驱动的商业智能文化数据可视化平台。为了实现这一目标，Microsoft Power BI 提供了自助式分析工具，可用于以有意义的方式分析、汇总和共享数据。

Microsoft Power BI 为客户提供数百种数据可视化以及内置的人工智能功能和 Excel 集成工具。它还提供多种支持系统，例如常见问题解答、论坛以及与工作人员的实时聊天。

9. Klipfolio

Klipfolio 是最好的数据可视化工具之一。你可以使用连接程序访问数百个不同数据源(例如电子表格、数据库、文件和 Web 服务应用程序)的数据。Klipfolio 还允许创建自定义的拖放数据可视化，你可以从中选择不同的选项，如图表、图形、散点图等。

Klipfolio 还有一些可以执行复杂公式的工具，从而解决具有挑战性的数据问题。如果有客户询问，可以从社区论坛或知识论坛获得帮助。

10. SAP Analytics Cloud

SAP Analytics Cloud 使用商业智能和数据分析功能来评估数据并创建可视化以预测业务成果。它还提供最新的建模工具，通过提醒数据中可能存在的错误并对不同的数据度量和维度进行分类来帮助用户。SAP Analytics Cloud 还建议对数据进行智能转换，从而增强可视化效果。

如果你对数据可视化有任何疑问或业务问题，SAP Analytics Cloud 将通过对话式人工智能和自然语言技术处理用户查询，确保用户满意。

Bigdata **Exercises**

〖Ex. 1〗 根据 Text A 回答以下问题。

1. What is data visualization?
2. What did businesses need as they accumulated massive collections of data during the early years of the big data trend?
3. What can data visualization also help businesses do?
4. What are companies increasingly using machine learning to do?
5. What does big data visualization require powerful computer systems to do?
6. Why do big data visualization projects often require involvement from IT as well as management?
7. What are some other popular techniques mentioned in the passage?
8. What does research from the media agency Magna predict?
9. What do common use cases for data visualization include?
10. What does scientific visualization allow scientists and researchers to do?

〖Ex. 2〗 根据 Text B 回答以下问题。

1. What is Tableau?
2. What does Looker provide?
3. What is Zoho Analytics?
4. What does Sisense believe?
5. Why can you use IBM Cognos Analytics easily even if you have limited or no knowledge about data analytics?
6. How can you deploy any combination of SaaS, on-premises or a private cloud?
7. What is Domo? What do these tools provide?
8. What does Microsoft Power BI offer?
9. What does Klipfolio also allow you to do?
10. What does SAP Analytics Cloud provide?

〖Ex. 3〗 把下列词组翻译成中文。

1. data analytics engine   1. _____
2. modeling tool   2. _____

3.	multiple user	3. _____
4.	data-driven company	4. _____
5.	smart assistant	5. _____
6.	visualization output	6. _____
7.	customer support	7. _____
8.	time series chart	8. _____
9.	pie chart	9. _____
10.	information visualization	10. _____

〖Ex. 4〗 把下列单词翻译成英文。

1.	v. 显示 n. 显示器	1. _____
2.	n. 分配，分布	2. _____
3.	v. 排除，清除	3. _____
4.	vt. 保证，担保	4. _____
5.	n. 直方图	5. _____
6.	n. 信息图	6. _____
7.	n. 机会，时机	7. _____
8.	n. 特性；属性	8. _____
9.	n. 结合(体)，联合(体)	9. _____
10.	vt. 安装	10. _____

〖Ex. 5〗 翻译句子。

1. Information visualization is a hot research problem in the information management field.
2. All kinds of proportion data in pie chart is called composition data.
3. An insecure computer system is an open door to criminals.
4. This information is only raw data and will need further analysis.
5. Graph can be a very useful tool for conveying information, especially numbers, percentages, and other data.
6. An error code will be displayed if any invalid information has been entered.
7. Access control creates multiple user accounts and authorizes each account to possess different access rights.
8. Follow the instructions below to download and install the patch.
9. It has a good features and customer support.
10. The community can share their ideas, suggestions and experiences via the community forum.

Top 9 Job Roles in the World of Data Science

New York Times considers Data Science as a "hot new field that promises to revolutionize industries from business to government, health care to academia[①]." However, there are a variety of different jobs and roles under the data science umbrella to choose from.

1. Data Analyst

Data analysts are responsible for[②] a variety of tasks including visualisation, munging[③], and processing of massive amounts of data. They also have to perform queries on the databases from time to time[④]. One of the most important skills of a data analyst is optimization. This is because they have to create and modify[⑤] algorithms that can be used to cull[⑥] information from some of the biggest databases without corrupting the data.

How to become a data analyst?

SQL, R, SAS, Python are some of the sought after technologies for data analysis. So, certification[⑦] in these can easily give a boost to your job applications. You should also have good problem-solving qualities.

2. Data Engineers

Data engineers[⑧] build and test scalable big data ecosystems for the businesses so that the data scientists can run their algorithms on the data systems that are stable and highly optimized. Data engineers also update the existing systems with newer or upgraded[⑨] versions of the current technologies to improve the efficiency of the databases.

How to become a data engineer?

If you are interested in a career as a data engineer, then technologies that require hands-on

---

① academia [ˌækəˈdiːmiə] n. 学术界

② be responsible for：负责

③ munging [ˈmʌŋgɪŋ] n. 整理

④ from time to time：有时；偶尔；间或

⑤ modify [ˈmɒdɪfaɪ] vi. 修改

⑥ cull [kʌl] vt. 挑选，精选；剔除

⑦ certification [ˌsɜːtɪfɪˈkeɪʃn] n. 证明，鉴定，认证

⑧ engineer [ˌendʒɪˈnɪə] n. 工程师

⑨ upgrade [ˌʌpˈgreɪd] vt. 提升；使(机器、计算机系统等)升级

experience include Hive, NoSQL, R, Ruby, Java, C++, and Matlab. It would also help if you can work with popular data APIs and ETL tools, etc.

3. Database Administrator

The job profile of a database administrator[①] is pretty much self-explanatory[②]—they are responsible for the proper functioning of all the databases of an enterprise and grant[③] or revoke[④] its services to the employees of the company depending on their requirements. They are also responsible for database backups and recoveries.

How to become a database administrator?

Some of the essential skills and talents of a database administrator include database backup and recovery, data security, data modelling, and design, etc. If you are good at disaster management[⑤], it's certainly a bonus.

4. Machine Learning Engineer

Machine learning engineers are in high demand today. However, the job profile comes with its challenges. Apart from having in-depth knowledge in some of the most powerful technologies such as SQL, REST[⑥] APIs, etc. machine learning engineers are also expected to perform A/B testing, build data pipelines, and implement common machine learning algorithms such as classification, clustering, etc.

How to become a machine learning engineer?

Firstly, you must have a sound knowledge of some of the technologies like Java, Python, JS, etc. Secondly, you should have a strong grasp[⑦] of statistics and mathematics. Once you have mastered both, it's a lot easier to crack a job interview.

5. Data Scientist

Data scientists have to understand the challenges of business and offer the best solutions using data analysis and data processing. For instance, they are expected to perform predictive analysis and run a fine-toothed comb[⑧] through an "unstructured/disorganized[⑨]" data to offer actionable insights. They can also do this by identifying trends and patterns that can help the

---

① administrator [əd'mɪnɪstreɪtə] n. 管理员，管理者
② self-explanatory [self ɪk'splænətri] adj. 不解自明的，明显的
③ grant [grɑːnt] v. 授予；承认；允许
④ revoke [rɪ'vəʊk] vt. 撤销，取消；废除
⑤ disaster management：灾难管理
⑥ REST(Representational State Transfer)：表现层状态转移
⑦ grasp [grɑːsp] n. 掌握；理解
⑧ fine-toothed comb：非常细致地
⑨ disorganized [dɪs'ɔːgənaɪzd] adj. 混乱的，杂乱无章的

companies in making better decisions.

How to become a data scientist?

To become a data scientist, you have to be an expert in R, MatLab, SQL, Python, and other complementary[①] technologies. It can also help if you have a higher degree in mathematics or computer engineering, etc.

6. Data Architect

A data architect[②] creates the blueprints[③] for data management so that the databases can be easily integrated, centralized, and protected with the best security measures. They also ensure that the data engineers have the best tools and systems to work with.

How to become a data architect?

A career in data architecture requires expertise in data warehousing, data modelling, extraction transformation and loan (ETL), etc. You also must be well versed[④] in Hive, Pig, and Spark, etc.

7. Statistician

A statistician[⑤], as the name suggests, has a sound understanding of statistical theories and data organization. Not only do they extract and offer valuable insights from the data clusters, but they also help create new methodologies for the engineers to apply.

How to Become a Statistician?

A statistician has to have a passion for logic. They are also good with a variety of database systems such as SQL, data mining, and the various machine learning technologies.

8. Business Analyst

The role of business analysts is slightly different from other data science jobs. While they do have a good understanding of how data-oriented technologies work and how to handle large volumes of data, they also separate the high-value data from the low-value data. In other words, they identify how the big data can be linked to actionable business insights for business growth.

How to become a business analyst?

Business analysts act as a link between the data engineers and the management executives. So, they should have an understanding of business finances and business intelligence, and also the IT technologies like data modelling, data visualization tools, etc.

---

① complementary [ˌkɒmplɪ'mentəri] adj. 互补的，相辅相成的
② architect ['ɑːkɪtekt] n. 架构师
③ blueprint ['bluːprɪnt] n. 蓝图，设想
④ be well versed：精通，熟练
⑤ statistician [ˌstætɪ'stɪʃn] n. 统计师，统计员

9. Data and Analytics Manager

A data and analytics manager oversees the data science operations and assigns the duties[①] to their team according to skills and expertise. Their strengths[②] should include technologies like SAS, R, SQL, etc. and of course management.

How to become a data and analytics manager?

First and foremost you must have excellent social skills[③], leadership qualities, and an out-of-box[④] thinking attitude. You should also be good at data science technologies like Python, SAS, R, Java, etc.

---

① duty ['djuːti]　n. 责任；职务，职责
② strength [streŋθ]　n. 优势；实力
③ social skill：社交技能
④ out-of-box：创造性的

Unit 9

Big Data Security

1. What Is Data Security?

Big data security is a general term used to describe all instruments and methods of guarding the data and analytics processes from attacks, being stolen or other foul play activities that could have a negative impact. Similar to other types of cyber security, big data attacks could either come from online or offline threats.

If you operate in the cloud, concerns stemming from securing big data are even greater. Such threats include theft of information stored online, ransomware, or DDoS attacks which could crash your server. If you store sensitive or classified information i.e. customer data, credit card numbers, or even simply contact details, a security breach could cost you even more. An attack on your big data storage could result in severe financial consequences such as monetary losses, court costs, fines or sanctions.

2. Big Data Security Technologies

2.1 Encryption

Encryption of data is generally done to secure a massive volume of data and different types of data. Data can be user-generated or machine-generated code. Encryption tools along with different analytics toolsets format or code the data. They also get applied to data from different sources like relational database management system (RDBMS), specialized file systems like Hadoop Distributed File System (HDFS), etc.

2.2 User access control

User access control is the most basic network security tool. Automated strong user access control is a must for organizations. Automation control manages complex user control levels and protects the big data platform against the inside attack (see Figure 9-1).

Figure 9-1　An Example of the UAC Consent Prompt

2.3　Physical security

Physical security should not be ignored. It is generally built in when you deploy the big data platform in your own center. It can also be built around your cloud provider's data center security.

Physical security is important as it can deny data center access to strangers or suspicious visitors. Video surveillance and security logs are also used for the same purpose.

2.4　Centralized key management

Centralized key management is one of the best security practices for many years. It is applied in big data environments, especially on those having wide geographical distribution. Best practices under centralized key management include policy-driven automation, on-demand key delivery, logging and abstracting key management from key usage.

3.　Big Data Security Use Cases

3.1　Cloud security monitoring

Cloud computing generally offers more efficient communication and increased profitability for all businesses. This communication needs to be secure.

Big data security offers cloud application monitoring. This provides host sensitive data and also monitors cloud-hosted infrastructure. Solutions also offer support across several relevant cloud platforms.

3.2　Network traffic analysis

Traffic continually moves in and out of your network. Due to the high volume of data over the network, it is difficult to maintain transactional visibility over the network traffic.

Security analytics allows your enterprise to watch over this network traffic. It is used to establish baselines and detect anomalies. This also helps in cloud security monitoring. It is used to analyze traffic in and out of cloud infrastructure. It also analyzes encrypted sensitive data, thus ensuring the proper working of channels.

3.3　Insider threat detection

Insider threats are as much as a danger to your enterprise as external threats. An active malicious user can do as much damage as any malware attack. But it is only in some rare cases that an insider threat can destroy a network.

With the help of security analytics, organizations can easily detect the insider threats. This is anticipated through behaviors such as abnormal login times, unusual email usage and unauthorized database access requests.

3.4 Threat hunting

Generally, the IT security team mostly engage in threat hunting. They search for potential indicators of dwelling threats and breaches that try to attack the IT infrastructure.

Security analytics helps to automate this threat of hunting. It acts as an extra set of eyes for your threat hunting efforts. Threats hunting automation can help in detecting malware beaconing activity and thus alerts for its stoppage as soon as possible.

3.5 Incident investigation

Generally, the sheer number of security alerts from SIEM solutions would overwhelm your IT security team. These continuous alerts can cause more burnout and frustration.

To minimize this issue, security analytics automates the incident investigation by providing contextualization to alerts. Thus your team has more time to prioritize incidents and can deal with potential breach incidents first.

3.6 User behaviour analysis

Users of your organization generally interact with your IT infrastructure all the time. Mainly it is the user's behavior that decides the success or failure of your cyber security. Therefore there is a need for tracking user's behavior.

The security analytics monitor the unusual behavior of employees. It helps to detect an insider threat or a malicious account. It can also detect suspicious patterns by correlating malicious activities.

3.7 Data exfiltration detection

Data exfiltration is termed as any unauthorized movement of data moving in and out of any network. Unauthorized data movements can cause theft and leakage of data.

Thus there is a need to protect data from such unauthorized access. Security analytics helps to detect the data exfiltration over a network. It is generally used to detect data leakage in encrypted communications.

4. Big Data Security Issues

• Access controls. It is critically important for an organization to have a system which is fully secure. Permission to exchange the data should be granted to authenticated users only. Access control needs to be such that it would not get hacked by attackers, hackers, or by any malicious activities. But to build a fully secure and strong access control is a big issue for organizations as it involves a big investment and a lot of maintenance.

• Non-relational data stores: Non-relational databases like NoSQL usually lack security by themselves.

• Storage: In big data architecture, we store data on multiple tiers. Its storage depends on business needs in terms of performance and cost. For example, high-priority data is generally stored on flash media. So locking down storage means creating a tier-conscious strategy.

• Endpoints: Security solutions that usually draw logs from endpoints will need to validate the authenticity of those endpoints or the analysis is not going to do much.

• Real-time security/compliance tools: Real-time tools generally generate a large amount of information. The key is to find a way to ignore false or rough information so that human talent can be focused on true breaches or valuable information.

• Data mining solutions: Data mining solutions generally find a pattern that suggests business strategies. For this reason, there is a need for ensuring that it is secured from both internal and external threats.

5. Some of the Typical Challenges in Securing Big Data

• It is difficult for security softwares to protect new toolsets or new technologies like advanced analytic tools for unstructured big data and non-relational databases like NoSQL.

• Security tools are very efficient and effective to protect data storage. But they do not have the same impact on data output from multiple analytical tools to multiple locations.

• Big data administrators generally mine the data without permission or notification. But whether the motivation is curiosity or criminal profit, security tools need to monitor only on suspicious access.

• If there is no regular update of security done by a big data owner, there is risk of data loss and exposure.

• Big data security experts need to continuously update their knowledge regarding cleanup and removal of malware and threats.

6. Conclusion

There are several challenges in securing big data. So when you are hosting the big data platform in the cloud, try to take nothing for granted. There is a need to work closely with the service provider to overcome these challenges with strong security SLA (Service Level Agreement).

Who is responsible for the security of this vital information? Everyone, yes, almost everyone working in an organization is responsible for securing the important data.

IT team is responsible for policies and procedures, and security software. Software security helps to protect big data deployment against malware and unauthorized access.

Compliance officers must work with the IT team closely to protect compliance. To safeguard the databases DBA (Database Administrators) should also work closely with the IT team.

Securing the big data platform from high and low threats will allow your organization to provide services well in the long run.

New Words

instrument	['ɪnstrəmənt]	n. 工具；仪器
attack	[ə'tæk]	v. 攻击
cyber	['saɪbə]	adj. 计算机(网络)的，信息技术的
online	[ˌɒn'laɪn]	adj. 在线的；联网的；联机的
offline	[ˌɒf'laɪn]	adj. 未连线的；脱机的；离线的
threat	[θret]	n. 威胁；征兆
concern	[kən'sɜːn]	n. 担心；令人担心的事
ransomware	['rænsəmweə]	n. 勒索软件
crash	[kræʃ]	v. 攻破；使(计算机)崩溃，瘫痪
sensitive	['sensətɪv]	adj. 敏感的
breach	[briːtʃ]	n. 违反，破坏；缺口
		v. 违反，破坏；突破，攻破
consequence	['kɒnsɪkwəns]	n. 结果，后果
monetary	['mʌnɪtəri]	adj. 货币的，金钱的
fine	[faɪn]	n. 罚款
		v. 对……处以罚金
sanction	['sæŋkʃn]	n. 制裁，处罚
encryption	[ɪn'krɪpʃn]	n. 加密
toolset	['tuːlset]	n. 成套工具，工具箱
provider	[prə'vaɪdə]	n. 供应者，提供者
deny	[dɪ'naɪ]	v. 否认；拒绝
stranger	['streɪndʒə]	n. 陌生人
suspicious	[sə'spɪʃəs]	adj. 可疑的，猜疑的，怀疑的
surveillance	[sɜː'veɪləns]	n. 监视
log	[lɒg]	n. 日志；记录
profitability	[ˌprɒfɪtə'bɪləti]	n. 收益性；利益率
host	[həʊst]	n. 主机
		v. 托管
baseline	['beɪslaɪn]	n. 基线
transactional	[træn'zækʃənəl]	adj. 业务的，交易的
visibility	[ˌvɪzə'bɪləti]	n. 可见性；清晰度

proper	['prɒpə]	adj. 正常的，正确的；适宜的
channel	['tʃænl]	n. 通道，渠道
danger	['deɪndʒə]	n. 危险，风险
malicious	[mə'lɪʃəs]	adj. 恶意的，有敌意的
malware	['mælweə]	n. 恶意软件
destroy	[dɪ'strɔɪ]	v. 毁坏，毁掉
detect	[dɪ'tekt]	v. 发现；查明；检测出
anticipate	[æn'tɪsɪpeɪt]	v. 预期；预计
abnormal	[æb'nɔːml]	adj. 异常的
email	['iːmeɪl]	n. 电子邮件
		v. 给……发电子邮件
unauthorize	[ʌn'ɔːθəraɪz]	v. 未批准，未授权
hunt	[hʌnt]	v. 搜寻；追踪；打猎
beacon	['biːkən]	n. 信标
activity	[æk'tɪvəti]	n. 活动，行动
burnout	['bɜːnaʊt]	n. 倦怠，筋疲力尽
frustration	[frʌ'streɪʃn]	n. 挫折；失败
incident	['ɪnsɪdənt]	n. 事件
investigation	[ɪnˌvestɪ'geɪʃn]	n. 调查
contextualization	[kənˌtekstʃuəlaɪ'zeɪʃn]	n. 情景化，语境化
prioritize	[praɪ'ɒrətaɪz]	v. 按重要性排列，划分优先顺序
exfiltration	[eksfɪl'treɪʃən]	n. 泄漏
leakage	['liːkɪdʒ]	n. 漏出，泄露
permission	[pə'mɪʃn]	n. 准许；许可证
exchange	[ɪks'tʃeɪndʒ]	v.&n. 交换；交流
grant	[graːnt]	v. 承认；允许
investment	[ɪn'vestmənt]	n. 投资，投入
endpoint	['endpɔɪt]	n. 端点
validate	['vælɪdeɪt]	v. 验证，证实，确认
ignore	[ɪg'nɔː]	v. 忽略，忽视
rough	[rʌf]	adj. 粗糙的，粗略的
talent	['tælənt]	n. 人才，天才；天资，天赋
valuable	['væljuəbl]	adj. 有价值的，宝贵的
notification	[ˌnəʊtɪfɪ'keɪʃn]	n. 通知；布告；公布
curiosity	[ˌkjʊərɪ'ɒsəti]	n. 好奇心
criminal	['krɪmɪnl]	n. 罪犯

		adj. 犯罪的
regular	['reɡjʊlə]	adj. 有规律的，定期的
cleanup	['kli:nʌp]	n. 清扫，清除

Phrases

big data security	大数据安全
negative impact	负面影响
credit card number	信用卡号码
result in	导致；造成
user access control	用户访问控制
physical security	物理安全，实际安全
centralized key management	集中式密钥管理
cloud-hosted infrastructure	云托管基础设施
traffic analysis	流量分析
cloud security monitoring	云安全监控
threat detection	威胁检测
external threat	外部威胁
insider threat	内部威胁
engage in	(使)从事；参与
deal with	处理，应付
data exfiltration detection	数据泄露检测
multiple tiers	多层
compliance tool	合规工具

Abbreviations

DDoS (Distributed Denial of Service)	分布式拒绝服务
RDBMS (Relational Database Management System)	关系数据库管理系统
HDFS (Hadoop Distributed File System)	Hadoop 分布式文件系统
SIEM (Security Information Event Management)	安全信息与事件管理
SLA (Service Level Agreement)	服务级别协议

 Analysis of Difficult Sentences

[1] Big data security is a general term used to describe all instruments and methods of

guarding the data and analytics processes from attacks, being stolen or other foul play activities that could have a negative impact.

本句中，used to describe all instruments and methods of guarding the data and analytics processes from attacks, being stolen or other foul play activities that could have a negative impact 是过去分词短语，作定语，修饰和限定 a general term。在该短语中，that could have a negative impact 是定语从句，修饰和限定 other foul play activities。

[2]　To minimize this issue, security analytics automates the incident investigation by providing contextualization to alerts.

本句中，To minimize this issue 是动词不定式短语，作目的状语，修饰谓语 automates。by providing contextualization to alerts 是介词短语，作方式状语，也修饰谓语 automates。

[3]　But to build a fully secure and strong access control is a big issue for organizations as it involves a big investment and a lot of maintenance.

本句中，to build a fully secure and strong access control 是动词不定式短语，作主语。as it involves a big investment and a lot of maintenance 是原因状语从句，修饰谓语 is a big issue for organizations。

[4]　The key is to find a way to ignore false or rough information so that human talent can be focused on true breaches or valuable information.

本句中，to find a way to ignore false or rough information so that human talent can be focused on true breaches or valuable information 是动词不定式短语，作表语。在该短语中，动词不定式短语 to ignore false or rough information 作定语，修饰和限定 a way。so that human talent can be focused on true breaches or valuable information 是目的状语从句，修饰 to find a way。

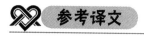 参考译文

大 数 据 安 全

1. 什么是数据安全？

大数据安全是一个通用术语，用于描述保护数据和分析过程免受攻击、被盗或其他可能产生负面影响的恶意活动的所有工具和方法。与其他类型的网络安全类似，大数据攻击可能来自在线威胁也可能来自离线威胁。

如果你在云中操作，那么保护大数据安全的问题就更为严峻。此类威胁包括盗窃在线存储的信息、使用勒索软件或进行可能导致服务器崩溃的 DDoS 攻击。如果你存储了敏感或机密信息，例如客户数据、信用卡号，甚至只是联系方式，安全漏洞也可能会让你付出更多代价。攻击你的大数据存储会导致严重的财务后果，例如金钱损失、法庭费用、罚款或制裁。

2. 大数据安全技术

2.1 加密

数据加密通常用于保护大量数据和不同类型的数据。数据可以是用户或机器生成的代码。加密工具以及不同的分析工具集对数据进行格式化或编码。它们还可以应用于不同来源的数据，如关系数据库管理系统(RDBMS)、专用文件管理系统[如 Hadoop 分布式文件系统(HDFS)]等。

2.2 用户访问控制

用户访问控制是最基本的网络安全工具。强大的自动用户访问控制是组织所必需的。自动化控制可以管理复杂的用户控制级别并保护大数据平台免受内部攻击(见图 9-1)。

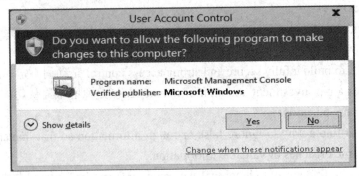

图 9-1 用户账户控制许可提示的例子

2.3 物理安全

物理安全不应被忽视。一般是在中心部署大数据平台的时候就要将其内置。它还可以围绕云提供商的数据中心安全性进行构建。

物理安全很重要，它可以拒绝陌生人或可疑访客访问数据中心。视频监控和安全日志也用于相同目的。

2.4 集中式密钥管理

集中式密钥管理是多年来最好的安全实践之一。它适用于大数据环境，特别是那些具有广泛地域分布的环境。集中式密钥管理下的最佳实践包括策略驱动的自动化、按需密钥交付、日志记录以及从密钥使用中提取密钥管理。

3. 大数据安全用例

3.1 云安全监控

云计算通常为所有企业提供更高效的通信和更强的盈利能力。这种通信必须安全。

大数据安全提供云应用监控。这样做可以保护主机的敏感数据并监控云托管基础设施。解决方案还支持跨多个相关云平台。

3.2 网络流量分析

数据流不断进出你的网络。由于网络上的数据量很大，很难保持对网络流量事务的可见性。

安全分析允许企业监视此网络流量。安全分析用于建立基线和检测异常，也有助于云安全监控。安全分析不仅用于分析进出云基础设施的流量，还分析加密的敏感数据，从而确保通道的正常工作。

3.3　内部威胁检测

内部威胁与外部威胁一样对企业构成威胁。活跃的恶意用户可以造成与任何恶意软件攻击同等的损害。但只有在极少数情况下，内部威胁才能破坏网络。

在安全分析的帮助下，企业可以轻松检测内部威胁。这是通过异常登录时间、异常电子邮件使用和未经授权的数据库访问请求等行为来预测的。

3.4　威胁搜寻

通常，IT 安全团队主要从事威胁搜寻。他们寻找潜在的威胁指标和试图攻击 IT 基础设施的漏洞。

安全分析有助于使威胁搜寻自动化。它可以充当威胁搜寻工作的额外眼睛。威胁搜寻自动化可以帮助检测恶意软件信标活动，从而尽快发出停止警报。

3.5　事故调查

通常，来自 SIEM 解决方案的大量安全警报会让 IT 安全团队不堪重负。这些持续的警报可能会导致团队更多的倦怠和沮丧。

为了最大限度地减少此问题，安全分析通过为警报提供情景来自动进行事件调查。因此，团队可以有更多时间来确定事件的优先级，并可以优先处理潜在的违规事件。

3.6　用户行为分析

企业的用户通常始终与 IT 基础架构交互。网络安全与否主要由用户的行为决定。因此需要跟踪用户的行为。

安全分析监控员工的异常行为。它有助于检测内部威胁或恶意账户，还可以通过关联恶意活动来检测可疑模式。

3.7　数据泄露检测

数据泄露指任何未经授权将数据移入或移出网络的行为。未经授权的数据移动可能导致数据被盗和泄露。

因此，需要保护数据免受此类未经授权的访问。安全分析有助于检测网络上的数据泄露。它通常用于检测加密通信中的数据泄漏。

4．大数据安全问题

* 访问控制。对于一个组织来说，拥有一个完全安全的系统至关重要。交换数据的权限应该只授予经过身份验证的用户。需要确保访问控制不会受到攻击者、黑客或任何恶意活动的攻击。但是建立一个完全安全和强大的访问控制对于组织来说是一个大问题，因为它会涉及大量投资和维护。

* 非关系型数据存储：NoSQL 等非关系型数据库通常本身缺乏安全性。

* 存储：在大数据架构中，我们将数据存储在多个层上。其存储取决于业务在性能和成本方面的需求。例如，高优先级数据通常存储在闪存介质上。因此，锁定存储意味着

创建一个有层次意识的策略。

• 端点：通常从端点提取日志的安全解决方案需要验证这些端点的真实性，否则分析将无济于事。

• 实时安全/合规工具：实时工具通常会生成大量信息。关键是找到一种方法来忽略虚假或粗略的信息，以便人才可以专注于真正的漏洞或有价值的信息。

• 数据挖掘解决方案：数据挖掘解决方案通常会找到一种建议业务策略的模式。因此，需要确保其免受内部和外部威胁。

5. 保护大数据的典型挑战

• 安全软件难以保护新工具集或新技术，例如用于非结构化大数据和非关系数据库(如 NoSQL)的高级分析工具。

• 安全工具能够非常有效地保护数据存储。但它们对从多个分析工具到多个位置的数据输出则无效。

• 大数据管理员通常在未经许可或通知的情况下挖掘数据。但是，无论其动机是出于好奇还是犯罪利益，安全工具只需要监控可疑访问。

• 如果大数据所有者没有定期更新安全性，则存在数据丢失和暴露的风险。

• 大数据安全专家需要不断更新自身关于清除和移除恶意软件和威胁的知识。

6. 结论

保护大数据有几个挑战。因此，当你在云中托管大数据平台时不能想当然。需要与服务提供商密切合作，通过强大的安全 SLA(服务级别协议)克服这些挑战。

谁负责这些重要信息的安全？每个人，是的，几乎每个在组织中工作的人都有责任确保重要数据的安全。

IT 团队负责政策和程序以及安全软件。软件安全有助于保护大数据部署免受恶意软件和未经授权的访问。

合规官必须与 IT 团队密切合作以保护合规性。为了保护数据库，DBA(数据库管理员)也应与 IT 团队密切合作。

从长远来看，保护大数据平台免受高危和低危威胁将使组织能够提供良好的服务。

Bigdata **Text B**

Data Protection and Data Privacy

The terms data protection and data privacy are often used interchangeably, but there is an important difference between the two. Data privacy defines who has access to data, while data protection provides tools and policies to actually restrict access to the data. Compliance regulations help ensure that user's privacy requests are carried out by companies, and companies are responsible to take measures to protect private user data.

1. What Is Data Protection?

Data protection is a set of strategies and processes you can use to secure the privacy, availability, and integrity of your data. It is sometimes also called data security or information privacy.

A data protection strategy is vital for any organization that collects, handles, or stores sensitive data. A successful strategy can help prevent data loss, theft, or corruption and can help minimize damage caused in the event of a breach or disaster.

Data protection solutions rely on technologies such as data loss prevention (DLP), storage with built-in data protection, firewalls, encryption, and endpoint protection.

2. What Is Data Privacy and Why Is it Important?

Data privacy is a guideline for how data should be collected or handled, based on its sensitivity and importance. Data privacy is typically applied to personal health information (PHI) and personally identifiable information (PII). This includes financial information, medical records, social security or ID numbers, names, birth dates, and contact information.

Data privacy concerns apply to all sensitive information that organizations handle, including that of customers, shareholders, and employees. Often, this information plays a vital role in business operations, development, and finances.

Data privacy helps ensure that sensitive data is only accessible to approved parties. It prevents criminals from being able to maliciously use data and helps ensure that organizations meet regulatory requirements.

Data privacy is enforced by data protection regulations. Non-compliance may result in monetary fines or loss of brand authority.

3. Data Protection vs. Data Privacy

Although both data protection and privacy are important and the two often come together, these terms do not represent the same thing.

3.1 One addresses policies, the other mechanisms

Data privacy is focused on defining who has access to data while data protection focuses on applying those restrictions. Data privacy defines the policies while data protection tools and processes.

Creating data privacy guidelines does not ensure that unauthorized users don't have access. Likewise, you can restrict access with data protections while still leaving sensitive data vulnerable. Both are needed to ensure that data remains secure.

3.2 Users control privacy, companies ensure protection

Another important distinction between privacy and protection is who is typically in control. For privacy, users can often control how much of their data is shared and with whom. For protection, it is up to the companies handling data to ensure that it remains private. Compliance regulations reflect this difference and are created to help ensure that users' privacy requests are enacted by companies.

4. Data Protection Technologies and Practices

When it comes to protecting your data, there are many storage and management options you can choose from. Solutions can help you restrict access, monitor activity, and respond to threats. Here are some of the most commonly used practices and technologies:

• Data loss prevention (DLP)—a set of strategies and tools that you can use to prevent data from being stolen, lost, or accidentally deleted. Data loss prevention solutions often include several tools to protect against and recover from data loss.

• Storage with built-in data protection—modern storage equipment provides built-in disk clustering and redundancy.

• Firewalls—utilities that enable you to monitor and filter network traffic. You can use firewalls to ensure that only authorized users are allowed to access or transfer data (see Figure 9-2).

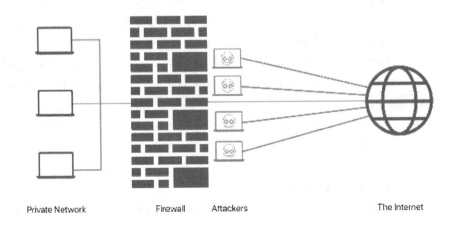

Private Network   Firewall   Attackers   The Internet

Figure 9-2　Firewalls

• Authentication and authorization—controls that help you verify credentials and assure that user privileges are applied correctly. These measures are typically used as part of an identity and access management (IAM) solution and in combination with role-based access controls (RBAC).

• Encryption—altering data content according to an algorithm that can only be reversed with the right encryption key. Encryption protects your data from unauthorized access even if data is stolen by making it unreadable.

• Endpoint protection—protecting gateways to your network, including ports, routers, and connected devices. Endpoint protection software typically enables you to monitor your network perimeter and to filter traffic as needed.

- Data erasure—limiting liability by deleting data that is no longer needed. This can be done after data is processed and analyzed or periodically when data is no longer relevant. Erasing unnecessary data is a requirement of many compliance regulations, such as GDPR.

5. Best Practices for Ensuring Data Privacy

Creating policies for data privacy can be challenging but it's not impossible. The following best practices can help you ensure that the policies you create are as effective as possible.

5.1 Practice minimal data collection

Make sure that your policies dictate that only necessary data is collected. If you collect more than what you need, you increase your liability and can create an undue burden on your security teams. Minimizing your data collection can also help you save on bandwidth and storage.

One way of achieving this is to use "verify not store" frameworks. These systems use third-party data to verify users and eliminate the need to store or transfer user data to your systems.

5.2 Include your users

Many users are aware of privacy concerns and are likely to appreciate transparency when it comes to how you're using and storing data. Therefore, GDPR has made user consent a key aspect of data use and collection.

You can be sure to include users and their consent in your processes by designing privacy concerns into your interfaces. For example, having clear user notifications outlining when data is collected and why. You should also include options for users to modify or opt out of data collection.

5.3 Inventory your data

Part of ensuring data privacy is understanding what data you have, how it is handled, and where it is stored. Your policies should define how this information is collected and acted upon. For example, you need to define how frequently data is scanned and how it is classified once located.

Your privacy policies should clearly outline what protections are needed for your various data privacy levels. Policies should also include processes for auditing protections to ensure that solutions are applied correctly.

5.4 Keeping up with data protection regulations

The widespread usage of personal and sensitive data has raised the significance of protecting this data from loss, and corruption. Global authorities have stepped in with regulatory compliance like General Data Protection Regulation (GDPR).

The GDPR emphasizes the personal data rights of EU residents, including the right to change, access, erase, or transfer their data. Personal data refers to any information that relates to an individual. This includes names, physical traits, addresses, racial or ethnic characteristics, and biometric data like DNA and fingerprints.

New Words

restrict	[rɪˈstrɪkt]	v. 限制，限定；约束
availability	[ə‚veɪləˈbɪləti]	n. 可用性；有效性
integrity	[ɪnˈtegrəti]	n. 完整性
vital	[ˈvaɪtl]	adj. 必要的，至关重要的
prevent	[prɪˈvent]	v. 防止，预防
corruption	[kəˈrʌpʃn]	n. 损坏
disaster	[dɪˈzɑːstə]	n. 灾难；不幸
guideline	[ˈgaɪdlaɪn]	n. 指导方针，指导原则
sensitivity	[‚sensəˈtɪvəti]	n. 敏感性
maliciously	[məˈlɪʃəsli]	adv. 有敌意地
mechanism	[ˈmekənɪzəm]	n. 机制，机能
vulnerable	[ˈvʌlnərəbl]	adj. 易受攻击的
distinction	[dɪˈstɪŋkʃn]	n. 区别
reflect	[rɪˈflekt]	v. 反映；反射
accidentally	[‚æksɪˈdentəli]	adv. 偶然地
recover	[rɪˈkʌvə]	v. 重新获得，恢复
redundancy	[rɪˈdʌndənsi]	n. 冗余；过多，过剩
utility	[juːˈtɪləti]	n. 实用程序；功用，效用
filter	[ˈfɪltə]	v. 过滤
		n. 筛选(过滤)程序
authentication	[ɔːˌθentɪˈkeɪʃn]	n. 身份验证；认证
credential	[krəˈdenʃl]	n. 凭证
		v. 提供证明书
privilege	[ˈprɪvəlɪdʒ]	n. 特权；优惠
		v. 特殊对待
reverse	[rɪˈvɜːs]	v. (使)反转；(使)颠倒；调换，交换
unreadable	[ʌnˈriːdəbl]	adj. 不能阅读的；难以理解的
gateway	[ˈgeɪtweɪ]	n. 门；入口；途径
port	[pɔːt]	n. 端口(计算机与其他设备的)接口
router	[ˈruːtə]	n. 路由器
perimeter	[pəˈrɪmɪtə]	n. 边界
erasure	[ɪˈreɪʒə]	n. 擦掉，删除
challenge	[ˈtʃælɪndʒ]	n. 挑战

impossible	[ɪmˈpɒsəbl]	adj. 不可能的
effective	[ɪˈfektɪv]	adj. 有效的
liability	[ˌlaɪəˈbɪləti]	n. 责任
undue	[ˌʌnˈdjuː]	adj. 过度的，过分的；不适当的
bandwidth	[ˈbændwɪdθ]	n. 带宽
appreciate	[əˈpriːʃɪeɪt]	v. 欣赏，重视
transparency	[trænsˈpærənsi]	n. 透明；透明度，透明性
consent	[kənˈsent]	v. 同意，允许
		n. 同意，许可；批文
modify	[ˈmɒdɪfaɪ]	v. 修改
inventory	[ˈɪnvəntəri]	v. 开列……的清单
outline	[ˈaʊtlaɪn]	v. 概述，略述

Phrases

data protection	数据保护
data privacy	数据隐私
rely on	依靠，依赖
storage equipment	存储设备
encryption key	密钥
opt out of ...	决定(从……)退出

Abbreviations

DLP (Data Loss Prevention)	数据丢失预防
PHI (Personal Health Information)	个人健康信息
PII (Personally Identifiable Information)	个人身份信息
IAM (Identity and Access Management)	身份和访问管理
RBAC (Role-Based Access Controls)	基于角色的访问控制

参考译文

数据保护和数据隐私

数据保护和数据隐私这两个术语经常互换使用，但两者之间有着重要区别。数据隐私定义了谁可以访问数据，而数据保护则提供了工具和政策来实际限制对数据的访问。合规性法规有助于确保用户的隐私请求由公司执行，公司有责任采取措施保护用户隐私数据。

1．什么是数据保护？

数据保护是一组策略和流程，可用于保护数据的隐私、可用性和完整性。有时也称为数据安全或信息隐私。

数据保护策略对于收集、处理或存储敏感数据的任何组织都至关重要。成功的策略有助于防止数据丢失、被盗或损坏，也有助于最大限度地减少在发生数据泄露或灾难时造成的损失。

数据保护解决方案依赖于数据丢失防护(DLP)、内置数据保护存储、防火墙、加密和端点保护等技术。

2．什么是数据隐私，为什么重要？

数据隐私是根据数据的敏感性和重要来收集或处理数据的指南。数据隐私通常适用于个人健康信息(PHI)和个人身份信息(PII)。这包括财务信息、医疗记录、社会保险号码或身份证号码、姓名、出生日期和联系信息。

数据隐私问题适用于组织处理所有敏感信息，包括客户、股东和员工的信息。通常，这些信息在业务运营、开发和财务中起着至关重要的作用。

数据隐私有助于确保敏感数据只能被批准方访问。它可以防止犯罪分子恶意使用数据，并有助于确保组织达到监管要求。

数据隐私保护由数据保护法规强制执行。不遵守规定可能会导致罚款或失去品牌威信。

3．数据保护与数据隐私

虽然数据保护和隐私都很重要，而且两者经常结合在一起，但这些术语意义不同。

3.1　一个涉及策略，另一个涉及机制

数据隐私侧重于定义谁可以访问数据，而数据保护侧重于应用这些限制。数据隐私定义策略。而数据保护定义工具和流程。

创建数据隐私准则并不能确保未经授权的用户没有访问权限。同样，在敏感数据容易受到攻击时你可以通过数据保护限制访问。这两者都是确保数据安全所需要的。

3.2　用户控制隐私，公司确保保护

隐私和保护之间的另一个重要区别通常是谁控制。隐私方面，用户通常可以控制共享数据量以及与谁共享。保护方面，则由处理数据的公司来确保数据的私密性。合规性法规反映了这种差异，旨在帮助确保用户的隐私请求由公司制定。

4．数据保护技术和实践

在保护数据方面，有许多存储和管理选项可供选择。解决方案可以帮助你限制访问、监控活动和应对威胁。以下是一些最常用的做法和技术：

• 数据丢失防护(DLP)——一组可用于防止数据被盗、丢失或意外删除的策略和工具。数据丢失防护解决方案通常包括几种工具来防止数据丢失并从中恢复数据。

- 具有内置数据保护功能的存储——现代存储设备提供内置磁盘集群和冗余。
- 防火墙——能够监控和过滤网络流量的实用程序。你可以使用防火墙来确保只有被授权用户才能访问或传输数据(见图9-2)。

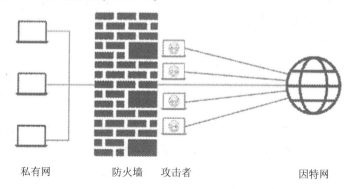

私有网　　　　　　防火墙　攻击者　　　　　　　　　因特网

图 9-2　防火墙

- 身份验证和授权——帮助你验证凭据并确保正确应用用户权限的控件。这些措施通常用作身份和访问管理(IAM)解决方案的一部分，并与基于角色的访问控制(RBAC)结合使用。
- 加密——根据只能用正确的加密密钥才能解密的算法来改变数据内容。即使数据被盗，加密也可以保护数据免遭未经授权的访问，使数据无法读取。
- 端点保护——保护网络的网关，包括端口、路由器和连接的设备。端点保护软件通常能够监控网络边界并根据需要过滤流量。
- 数据擦除——通过删除不再需要的数据来限制责任。这可以在数据处理和分析之后进行，也可以在数据不再相关时定期完成。删除不必要的数据是许多合规性法规(例如GDPR)的要求。

5. 确保数据隐私的最佳实践

制定数据隐私策略可能具有挑战性，但并非没有可能性。以下最佳实践有助于确保你所制定的策略尽可能有效。

5.1　尽量少地收集数据

确保策略规定只收集必要的数据。如果收集的数量超出了需要，则会增加责任，并可能给安全团队带来不必要的负担。最大限度地减少数据收集还有助于节省带宽和存储空间。

实现这一目标的一种方法是使用"验证而不存储"框架。这些系统使用第三方数据来验证用户，而无需将用户数据存储或传输到你的系统。

5.2　涵盖用户

许多用户都意识到了隐私问题，并且在涉及如何使用和存储数据时可能会偏好透明度。因此，GDPR已将用户同意作为数据使用和收集的关键。

通过将隐私问题设计到界面中，可以确保流程中包含用户及其同意。例如，有明确的用户通知，概述收集数据的时间和原因。还应该包括供用户修改或选择退出数据收集的选项。

5.3 清点数据

确保数据隐私的一部分是了解你拥有哪些数据、如何处理这些数据以及数据存储在什么地方。你的策略应该定义如何收集和处理这些信息。例如，需要定义扫描数据的频率以及定位后的分类方式。

隐私策略应清楚地概述各种数据隐私级别所需的保护措施。策略还应包括审核保护措施的流程，以确保正确应用解决方案。

5.4 遵守数据保护法规

个人和敏感数据的广泛使用提高了保护这些数据免遭丢失和损坏的重要性。全球权威机构已介入监管合规性，例如通用数据保护条例(GDPR)。

GDPR 强调欧盟居民的个人数据权利，包括更改、访问、删除或传输其数据的权利。个人数据是指与个人有关的任何信息，其中包括姓名、身体特征、地址、种族或民族特征，以及 DNA 和指纹等生物特征数据。

Bigdata **Exercises**

〖Ex. 1〗 根据 Text A 回答以下问题。

1. What is the purpose of encryption of data?
2. Why is physical security important?
3. What do best practices under centralized key management include?
4. What can organizations with the help of security analytics? How is this anticipated?
5. What can threats hunting automation do?
6. What mainly decides the success or failure of your cyber security?
7. What is data exfiltration termed as?
8. What do data mining solutions generally do?
9. What do big data administrators generally do?
10. What do big data security experts need to do??

〖Ex. 2〗 根据 Text B 回答以下问题。

1. What is data protection?
2. What do data protection solutions rely on?
3. What is data privacy?
4. What does data privacy do?
5. What is data privacy focused on?
6. What is DLP?
7. What can you use firewalls to do?
8. What does encryption do?

9. How can you be sure to include users and their consent in your processes?

10. What should your privacy policies do?

〖Ex. 3〗 把下列词组翻译成中文。

1.	big data security	1.	大数据安全
2.	data exfiltration detection	2.	数据泄露检测
3.	deal with	3.	处理，应付
4.	external threat	4.	外部威胁
5.	physical security	5.	物理安全，实际安全
6.	threat detection	6.	威胁检测
7.	user access control	7.	用户访问控制
8.	data privacy	8.	数据隐私
9.	traffic analysis	9.	流量分析
10.	data protection	10.	数据保护

〖Ex. 4〗 把下列单词翻译成英文。

1.	n. 身份验证，认证	1.	authentication
2.	n. 活动，行动	2.	activity
3.	adj. 异常的	3.	abnormal
4.	v. 攻击	4.	attack
5.	n. 违反，破坏；缺口	5.	breach
6.	v. 攻破；使(计算机)崩溃	6.	crash
7.	adj. 计算机(网络)的	7.	cyber
8.	v. 发现；查明；检测出	8.	detect
9.	n. 加密	9.	encryption
10.	v. &n. 交换；交流	10.	exchange

〖Ex. 5〗 翻译句子。

1. This module uses dual user access control measures to improve the security of the data in the system.

2. This passage is mainly about application of network traffic analysis technology and comparison of solutions.

3. DDoS attack is among the hardest network security problems to address.

4. DB2 is a highly powerful relational database management system with capabilities of storing and managing high volumes of data.

5. Security information and event management technologies can be used to improve user and role management.

6. A means of ensuring data protection should be established for all computerized system.

7. Experimental results show that the method can achieve better data privacy protection in the data mining.

8. You can change the encryption key for a database.

9. As an important component of the operating system, file system determines user data in storage equipment.

10. Comprehensive identity and access management is a vital criterion for success in the hybrid cloud.

Bigdata **Reading Material**

Big Data Privacy

1. What Is Big Data Privacy?

Big data privacy involves properly managing big data to minimize risk and protect sensitive data①. Because big data comprises② large and complex data sets, many traditional privacy processes cannot handle the scale and velocity required. To safeguard big data and ensure it can be used for analytics, you need to create a framework for privacy protection that can handle the volume, velocity, variety, and value of big data as it is moved between environments, processed, analyzed, and shared.

2. Big Data Includes Big Privacy Concerns

In an era of multi-cloud computing, data owners must keep up with both the pace of data growth and the proliferation③ of regulations that govern it—especially regulations protecting the privacy of sensitive data and personally identifiable information (PII④). With more data spread across more locations, the business risk of a privacy breach has never been higher, and with it, consequences⑤ ranges from high fines to loss of market share.

---

① sensitive data：敏感数据

② comprise [kəm'praɪz] vt. 包含，包括；由……组成；由……构成

③ proliferation [prə,lɪfə'reɪʃn] n. 增殖；激增

④ PII：个人识别信息

⑤ consequence ['kɒnsɪkwəns] n. 结果；重要性

Big data privacy is also a matter of customer trust. The more data you collect about users, the easier it gets to "connect the dots": to understand their current behavior, draw inferences about their future behavior, and eventually develop deep and detailed profiles of their lives and preferences. The more data you collect, the more important it is to be transparent① with your customers about what you're doing with their data, how you're storing it, and what steps you're taking to comply with regulations② that govern privacy and data protection.

The volume and velocity of data from existing sources, such as legacy applications and e-commerce, is expanding fast. You also have new (and growing) varieties of data types and sources, such as social networks and IoT device streams. To keep pace, your big data privacy strategy needs to expand, too. That requires you to consider all of these issues:

- What do you intend to do with customer and user data?
- How accurate is the data, and what are the potential consequences of inaccuracies③?
- How will your data security scale to keep up with threats of data breaches and insider threats as they become more common?
- Where is your balancing point④ between the need to keep data locked down in place and the need to expose it safely so you can extract value from it?
- How do you maintain compliance with data privacy regulations that vary across the countries and regions where you do business, and how does that change based on the type or origin of the data?
- How do you maintain transparency about what you do with the big data you collect without giving away the "secret sauce⑤" of the analytics that drive your competitive advantage?

3. Predictions for Big Data Privacy: What To Expect?

Prediction 1: Data privacy mandates will become more common.

As organizations store more types of sensitive data in larger amounts over longer periods of time, they will be under increasing pressure⑥ to be transparent about what data they collect, how they analyze and use it, and why they need to retain it. The European Union's General Data Protection Regulation (GDPR⑦) is a high-profile example. More government agencies and regulatory organizations are following suit. To respond to these growing demands, companies need reliable, scalable big data privacy tools that encourage⑧ and help people to access, review,

---

① transparent [træns'pærənt]　adj. 透明的；易懂的

② regulation [ˌregju'leɪʃn]　n. 规章；规则

③ inaccuracy [ɪn'ækjərəsi]　n. 不准确，误差

④ balancing point：平衡点

⑤ secret sauce：秘籍

⑥ pressure ['preʃə]　n. 压力

⑦ GDPR：通用隐私保护条例

⑧ encourage [ɪn'kʌrɪdʒ]　v. 鼓励；劝说；促进

correct, anonymize[①], and even purge some or all of their personal and sensitive information.

Prediction 2: New big data analytic tools will enable organizations to perform deeper analysis of legacy data, discover uses for which the data wasn't originally intended, and combine it with new data sources.

Big data analytics tools and solutions can now dig into data sources that were previously unavailable[②], and identify new relationships hidden in legacy data. That's a great advantage when it comes to getting a complete view of your enterprise data. But it also raises questions about the accuracy of aging data and the ability to track down entities for consent[③] to use their information in new ways.

The key to protecting the privacy of your big data while still optimizing its value is ongoing review of four critical data management activities:

- Data collection
- Retention[④] and archiving
- Data use, including use in testing, DevOps, and other data masking[⑤] scenarios
- Creating and updating disclosure[⑥] policies and practices

Companies with a strong, scalable data governance program will have an advantage when assessing these tasks—they will be able to accurately assess data-related risks and benefits in less time and quickly take more decisive[⑦] action based on trusted data.

4. Developing Big Data Privacy And Protection Strategies

Traditional data security is network-and-system-centric, but today's hybrid and multi-cloud architectures spread data across more platform-agnostic locations and incorporate more data types than ever before. Big data privacy can't be an afterthought[⑧]. It must be an integral[⑨] part of your cloud integration and data management strategy:

- You must define and manage data governance policies to clarify what data is critical and why, who owns the critical data, and how it can be used responsibly[⑩].
- You must discover, classify, and understand a wide range of sensitive data across all big data platforms at massive scale by leveraging artificial intelligence and machine learning tools to

---

① anonymize [ə'nɒnɪmaɪz]　vt. 使匿名，隐去姓名资料

② unavailable [ˌʌnə'veɪləbl]　adj. 难以获得的；不能利用的

③ consent [kən'sent]　v. 同意，允许

④ retention [rɪ'tenʃn]　n. 保留

⑤ data masking：数据屏蔽

⑥ disclosure [dɪs'kləʊʒə]　n. 公开；泄露

⑦ decisive [dɪ'saɪsɪv]　adj. 决定性的；明确的；决断的

⑧ afterthought ['ɑːftəθɔːt]　n. 事后的考虑或想法

⑨ integral ['ɪntɪɡrəl]　adj. 基本的；必需的

⑩ responsibly [rɪ'spɒnsəbli]　adv. 负责地

automate controls; then, you can use that information to develop and implement intelligent big data management policies.

• You must index, inventory, and link data subjects and identities to support data access rights and notifications.

• You must be able to perform continuous risk analysis for sensitive data to understand your risk exposure, prioritize① available data protection resources and investments, and develop protection and remediation② plans as your big data grows.

• You need automated, centralized big data privacy tools that integrate with native big data tools like Cloudera Sentry, Amazon Macie, and Hortonworks Ranger to streamline and facilitate the process of managing data access, such as viewing, changing, and adding access policies.

• You need fast and efficient data protection capabilities at scale, including dynamic masking for big data as it's put into use in production and data lakes, encryption for big data at rest in data lakes and data warehouses, and persistent masking for big data used in non-production environments like development and analytics.

• You must measure and communicate the status of big data privacy risk indicators as a critical part of tracking success in protecting sensitive information while supporting audit③ readiness.

---

① prioritize [praɪ'ɒrətaɪz]　vt. 按重要性排列，优先处理
② remediation [rɪˌmiːdɪ'eɪʃn]　n. 修复，补救，纠正
③ audit ['ɔːdɪt]　n. & v. 审计

Unit 10

Bigdata **Text A**

10 Big Data Trends

The vast changes prevailing in today's technological landscape has opened doors for big data to improve businesses across industries and boost economies. Its role has been elevated to a degree that extracting value from collected information has proven to be invaluable to companies both small and large. Big data helps organizations that have critical responsibilities in making the world a better place do their jobs better.

Big data wasn't as "big" during its early days. At that time only large-scale businesses were able to utilize it because they were the only ones that could afford the technology. Furthermore, the broad scope of their service needed more precise data, which calls for the need for a data analytics system.

Since then, big data has evolved at an incredibly fast rate. This has allowed even small businesses to benefit from it because of the cloud technology and the Internet. Big data cloud eliminates the need for an elaborate setup and expensive data experts since all the information they need can now be accessed remotely using an internet connection.

With the help of artificial intelligence, the cloud and the internet of things, even the complexity of big data can be handled by those who are willing to use it to their organization's advantage. Big data analytics has gone beyond the hot IT trend tag and has now established itself as part of doing business for companies. It is set to soon replace gold as one of the most valuable assets to man.

To keep you up-to-date, check out the hottest big data trends set to propel industries into the future.

1. Rapidly Growing IoT Networks

It's becoming quite common that our smartphones are being used to control our home appliances, thanks to the technology called the internet of things (IoT). With smart devices such as Google Assistant and Microsoft Cortana trending in homes to automate specific tasks, the growing IoT craze is drawing companies to invest in the technology's development (see Figure 10-1).

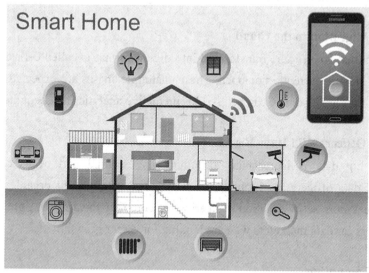

Figure 10-1   Smart Home

More organizations will jump on the opportunity in providing better IoT solutions. This will lead to more ways to collect vast amounts of data, and along with it the means to manage and analyze it. The industry response is to push for more new devices that are more capable of collecting, analyzing and processing data.

2. Accessible Artificial Intelligence

Artificial intelligence is now more commonly utilized to help both big and small companies improve their business processes. AI programs can now execute tasks much faster than humans, cut down errors along the way and improve the overall flow. This allows humans to focus better on more critical tasks and further enhance the quality of service.

The good news is everybody can have access to prebuilt machines that run AI applications to address the growing demand, which levels the playing field for companies in the same industry. Individual organizations may gain an advantage if they find the most efficient way to integrate this into their business process.

3. The Rise of Predictive Analytics

Big data analytics has always been a key strategy for businesses to have a competitive edge and achieve their goals. They use the necessary analytics tools to process big data and determine the reasons why certain events happen. Now, predictive analysis through big data can help predict what may occur in the future.

There is no doubt this kind of strategy is highly effective in helping analyze gathered information to predict consumer behavior. This allows companies to determine the measures they have to take by knowing a customer's next action. Analytics can also provide more context on data to help understand the reasons behind them.

4. Dark Data Migration to the Cloud

Information that is yet to be transformed into digital format is called dark data, and it is a huge "reservoir" that is currently untapped. These analog databases are expected to be digitized and migrated to the cloud, so they can be used for predictive analytics that benefits businesses.

5. Chief Data Officers Will Have Bigger Roles

Now that big data is increasingly becoming an essential part of executing business strategies, chief data officers are adopting a more critical role in their organization. They are expected to take a more active position in steering the company towards the right direction. This trend opens doors for data marketers who are looking for career growth.

6. Quantum Computing

Getting to analyze and interpret massive amounts of data can take a lot of time with the current technology we are using. If we can crunch billions of data at once in just a few minutes, we can cut processing time immensely, giving companies the opportunity to make timely decisions to achieve more desired results.

This huge undertaking can only be possible through quantum computing. Despite being in its infancy, experiments are currently being carried out on quantum computers in an effort to help in practical and theoretical research across different industries. Pretty soon, large tech companies such as Google, IBM, and Microsoft will start testing quantum computers to integrate them into their business processes.

7. Smarter and Tighter Cyber Security

Organizations have grown paranoid over the past scandals that involved hacking and system breaches. This has prompted them to focus on strengthening information confidentiality. IoT is also being a cause for concern with all the data being collected; cyber security is an issue. To address this perpetually impending threat, big data companies pitch in to help organizations use data analytics as a tool to predict and detect cyber security threats.

Big data can be integrated into a cyber security strategy through security log data where it can be used to provide information about past threats. This can help companies prevent and mitigate the impact of future hacks and data breaches.

8. Open Source Solutions

There are many public data solutions available, such as open source software, that have been making considerable improvements to speed up data processing. They now have features that allow access and response to data in real time. For this reason, they are expected to flourish and be in high demand.

There is no doubt open source software is cheaper in that it can help your business cut operations costs. However, there are some downsides that you need to know if you are willing to give them a shot.

9. Edge Computing

Edge computing is set to leave the cloud in the dust when it comes to processing data. Due to the growing trend of IoT, many companies are turning to connected devices to collect more data on customers or processes. This has created a need for technological innovations aimed to cut down on lag time from the gathering of data into the cloud, its analysis, and the action that needs to be taken.

Edge computing delivers a better performance since there is less data flowing in and out of the network, with less cloud computing costs. The company can also benefit from storage and infrastructure costs if they choose to delete unnecessary data collected from IoT. Additionally, edge computing can speed up data analysis, giving companies ample time to react.

10. Smarter Chatbots

Powered by smarter AI, chatbots are now being deployed by companies to handle customer queries to deliver more personalized interactions while eliminating the need for actual human personnel.

Big data has a lot to do with delivering a more pleasant customer experience as bots process large amounts of data to provide relevant answers based on the keywords entered by customers in their queries. During interactions, they are also able to collect and analyze information about customers from conversations. This process can help marketers develop a more streamlined strategy to achieve better conversions.

New Words

vast	[vɑːst]	adj. 巨大的，大量的；广阔的
prevail	[prɪ'veɪl]	v. 流行，盛行
large-scale	[lɑːdʒ skeɪl]	adj. 大规模的，大范围的
afford	[ə'fɔːd]	v. 负担得起，买得起
incredibly	[ɪn'kredəbli]	adv. 难以置信地；很，极为
Internet	['ɪntənet]	n. 互联网

elaborate	[ɪˈlæbərət]	adj. 复杂的；详尽的
remotely	[rɪˈməʊtli]	adv. 远程地，遥远地
establish	[ɪˈstæblɪʃ]	v. 创建；建立
propel	[prəˈpel]	v. 推进，推动；驱动，驱使
common	[ˈkɒmən]	adj. 常见的；共同的；普通的
response	[rɪˈspɒns]	n. 反应
enhance	[ɪnˈhɑːns]	v. 提高，增强；改进
prebuilt	[preˈbɪlt]	adj. 预建的，预制的
achieve	[əˈtʃiːv]	v. 实现，达到
determine	[dɪˈtɜːmɪn]	v. 查明；测定
certain	[ˈsɜːtn]	adj. 肯定的，确定的
event	[ɪˈvent]	n. 事件；活动
occur	[əˈkɜː]	v. 发生；存在于
migration	[maɪˈgreɪʃn]	n. 迁移，移居
untapped	[ˌʌnˈtæpt]	adj. 未开发的，未利用的
position	[pəˈzɪʃn]	n. 位置；立场
quantum	[ˈkwɒntəm]	n. 量子
crunch	[krʌntʃ]	v. (用计算机等大量)处理(数字)
infancy	[ˈɪnfənsi]	n. 初期，摇篮时代
tight	[taɪt]	adj. 牢固的
paranoid	[ˈpærənɔɪd]	adj. 极端疑惧的，多疑的，恐惧的
scandal	[ˈskændl]	n. 丑闻
strengthen	[ˈstreŋθn]	v. 巩固；支持；壮大；加强
confidentiality	[ˌkɒnfɪˌdenʃɪˈæləti]	n. 机密性
perpetually	[pəˈpetʃʊəli]	adv. 永恒地；终身地；不断地
impending	[ɪmˈpendɪŋ]	adj. 即将发生的，迫在眉睫的
mitigate	[ˈmɪtɪgeɪt]	v. 缓和，减轻
considerable	[kənˈsɪdərəbl]	adj. 相当大的，相当多的
flourish	[ˈflʌrɪʃ]	v. 繁荣；蓬勃发展
ample	[ˈæmpl]	adj. 大量的；充裕的
react	[riˈækt]	v. (作出)反应
chatbot	[tʃætbɒt]	n. 聊天机器人
interaction	[ˌɪntərˈækʃn]	n. 互动；交流，沟通
pleasant	[ˈpleznt]	adj. 令人愉快的；友好的
bot	[bɒt]	n. 机器人

| conversation | [ˌkɒnvəˈseɪʃn] | n. 交谈，谈话 |
| marketer | [ˈmɑːkɪtə] | n. 市场营销人员 |

Phrases

technological landscape	技术景观，技术格局
cloud technology	云技术
along with …	和……一起，随着
competitive edge	竞争优势
analytic tool	分析工具
consumer behavior	消费者行为
dark data	暗数据
chief data officer	首席数据官
critical role	关键角色，关键作用
quantum computing	量子计算
pretty soon	不久，很快
pitch in	作出贡献；参与；协力
log data	日志数据
open source software	开源软件
speed up	(使)加速；增速
operations cost	运营成本
edge computing	边缘计算
technological innovation	技术革新
lag time	延迟时间，滞后时间
data flow	数据流
computing cost	计算成本
infrastructure cost	基础设施成本
customer experience	客户体验

Analysis of Difficult Sentences

[1] Its role has been elevated to a degree that extracting value from collected information has proven to be invaluable to companies both small and large.

本句中，that extracting value from collected information has proven to be invaluable to companies both small and large 是定语从句，修饰和限定 a degree。both small and large 是形容词后置，作定语，修饰和限定 companies。

[2] Big data cloud eliminates the need for an elaborate setup and expensive data experts since

all the information they need can now be accessed remotely using an internet connection.

　　本句中，since all the information they need can now be accessed remotely using an internet connection 是原因状语从句，修饰谓语 eliminates。在该从句中，they need 是定语从句，修饰和限定 all the information。using an internet connection 是现在分词短语，作方式状语，修饰谓语 can now be accessed。

[3] The good news is everybody can have access to prebuilt machines that run AI applications to address the growing demand, which levels the playing field for companies in the same industry.

　　本句中，everybody can have access to prebuilt machines that run AI applications to address the growing demand, which levels the playing field for companies in the same industry 是表语从句。在该从句中，that run AI applications 是定语从句，修饰和限定 prebuilt machines。to address the growing demand 是动词不定式短语，作目的状语，修饰 can have access to。which levels the playing field for companies in the same industry 是非限定性定语从句，对 everybody can have access to prebuilt machines that run AI applications to address the growing demand 进行补充说明。

[4] Big data has a lot to do with delivering a more pleasant customer experience as bots process large amounts of data to provide relevant answers based on the keywords entered by customers in their queries.

　　本句中，as bots process large amounts of data to provide relevant answers based on the entered keywords by customers in their queries 是原因状语从句，修饰谓语 has a lot to do。to provide relevant answers based on the entered keywords by customers in their queries 是动词不定式短语，作目的状语，修饰 process large amounts of data。在该短语中，based on the keywords entered by customers in their queries 是过去分词短语作状语，修饰 provide relevant answers。过去分词短语 entered by customers in their queries 作定语，修饰 the keywords。based on 的意思是"根据，基于"。

参考译文

大数据十大趋势

　　当今技术领域中普遍存在的巨大变化为大数据改善各行业的业务并促进经济发展打开了大门。大数据的作用已经提升到一定程度，从收集的信息中提取价值，已被证明对大小公司都是极具价值的。大数据可以帮助那些在让世界变得更美好的方面负有关键责任的组织更好地完成工作。

　　大数据在其早期并没有那么"大"。当时，只有大型企业才能使用，因为只有大公司才能负担得起这项技术。此外，大公司的服务范围广泛，需要更精确的数据，这就需要数据分析系统。

从那时起，大数据以惊人的速度发展。由于云技术和互联网的发展，即使是小型企业也可以从中受益。大数据云消除了对精心设置和昂贵的数据专家的需要，因为现在可以通过互联网连接远程访问他们需要的所有信息。

在人工智能、云和物联网的帮助下，甚至大数据的复杂性也可以由那些愿意使用它来为组织带来优势的人处理。大数据分析已经成为超越热门的 IT 趋势标签，现在已经是公司开展业务的一部分。它将很快取代黄金，成为人类最宝贵的资产之一。

为了及时了解最新信息，下面介绍未来具有行业推动性的最热门的大数据趋势。

1. 快速增长的物联网

得益于物联网(IoT)技术，用智能手机控制家用电器越来越普遍。随着谷歌助手和微软 Cortana 等智能设备在家庭中实现特定任务的自动化，日益增长的物联网热潮正在吸引公司投资于该技术的开发(见图 10-1)。

更多的组织将抓住机会提供更好的物联网解决方案。这将催生更多收集大量数据的方法，以及管理和分析数据的方法。业界的反应是推动更多能够收集、分析和处理数据的新设备。

图 10-1　智能家居

2. 无障碍人工智能

人工智能现在更常用于帮助大小型公司改进其业务流程。人工智能程序现在可以比人类更快地执行任务，减少过程中的错误并改善整体流程。这使人类能够更好地专注于更关键的任务，并进一步提高服务质量。

好消息是每个人都可以使用运行 AI 应用程序的预制机器来满足不断增长的需求，这为同行业的公司提供了公平的竞争环境。如果单个组织找到最有效的方法将其集成到其业务流程中，就可能获得优势。

3．预测分析的兴起

大数据分析一直是企业获得竞争优势和实现目标的关键策略。这些企业使用必要的分析工具来处理大数据并确定某些事件发生的原因。现在，通过大数据进行预测分析就可以帮助预测未来可能发生的事情。

毫无疑问，这种策略在帮助分析收集的信息以预测消费者行为方面非常有效。这使公司能够通过了解客户的下一步行动来确定公司必须采取的措施。分析还可以提供有关数据的更多背景，以帮助了解其背后的原因。

4．暗数据上云

尚未转化为数字格式的信息称为暗数据，是目前尚未开发的巨大"水库"。这些模拟数据库预计将被数字化并迁移到云中，因此它们可用于预测分析从而为企业带来益处。

5．首席数据官将扮演更重要的角色

如今大数据越来越成为执行业务策略的重要组成部分，首席数据官在组织中扮演着更重要的角色。首席数据官将采取更积极的立场，引导公司朝着正确的方向发展。这一趋势为寻求职业发展的数据营销人员打开了大门。

6．量子计算

使用现有技术分析和解释海量数据可能需要花费大量时间。如果能在几分钟内一次性处理数十亿条数据，就可以极大地缩短处理时间，让公司有机会及时做出决策以实现更理想的结果。

这项艰巨的任务只能通过量子计算来实现。尽管量子计算还处于起步阶段，但目前正在量子计算机上进行实验，以帮助不同行业的实践和理论研究。很快，谷歌、IBM 和微软等大型科技公司将开始测试量子计算机，将它们集成到自身的业务流程中。

7．更智能、更严密的网络安全

组织对过往涉及黑客攻击和系统破坏的各种丑闻变得越来越疑惧。这促使他们专注于加强信息保密。物联网也引起人们对收集的所有数据的担忧；网络安全是一个问题。为了应对这种永远迫在眉睫的威胁，大数据公司纷纷出手帮助组织将数据分析用作预测和检测网络安全威胁的工具。

大数据可以通过将安全日志数据集成到网络安全策略中，提供有关过去威胁的信息。这可以帮助公司预防和减轻未来黑客攻击和数据泄露的影响。

8．开源解决方案

目前有许多可用的公共数据解决方案，例如开源软件，它们已经进行了巨大的改进以加速数据处理。开源软件现在具有允许实时访问和响应数据的功能。因此，开源软件有望

蓬勃发展并且需求量很大。

毫无疑问，开源软件更便宜，因为它可以帮助企业降低运营成本。但是，如果你有意尝试开源软件，就需要了解它的缺点。

9. 边缘计算

在处理数据时，边缘计算可以轻松击败云。由于物联网的发展，许多公司正在转向连接设备来收集有关客户或流程的更多数据。这就产生了对技术创新的需求，旨在减少从数据收集到云、分析和需要采取的行动的滞后时间。

因为流入和流出网络的数据更少，云计算成本更低，边缘计算提供了更好的性能。如果公司选择删除从物联网收集的不必要的数据，还可以从存储和基础设施成本中受益。此外，边缘计算可以加速数据分析，让公司有足够的时间做出反应。

10. 更智能的聊天机器人

在人工智能的支持下，很多公司正在部署聊天机器人来处理客户查询，以提供更个性化的交互，同时减少对实际人员的需求。

大数据与提供更愉快的客户体验之间存在很大关系，因为机器人处理大量数据，根据客户在查询中输入的关键字提供相关答案。在互动过程中，机器人还能够从对话中收集和分析有关客户的信息。此过程可以帮助营销人员制定更简化的策略以实现更好的转化。

Bigdata Text B

Big Data in the Cloud

1. The Pros of Big Data in The Cloud

The cloud brings a variety of important benefits to businesses of all sizes. Some of the most immediate and substantial benefits of big data in the cloud include the following.

1.1 Scalability

A typical business data center faces limits in physical space, power, cooling and the budget to purchase and deploy the sheer volume of hardware it needs to build a big data infrastructure. By comparison, a public cloud manages hundreds of thousands of servers spread across a fleet of global data centers. The infrastructure and software services are already there, and users can assemble the infrastructure for a big data project of almost any size.

1.2 Agility

Not all big data projects are the same. One project may need 100 servers, and another project might demand 2,000 servers. With cloud, users can employ as many resources as needed to accomplish a task and then release those resources when the task is complete.

1.3　Cost

A business data center is an enormous capital expense. Beyond hardware, businesses must also pay for facilities, power, ongoing maintenance and more. The cloud works all those costs into a flexible rental model where resources and services are available on demand and follow a pay-per-use model.

1.4　Accessibility

Many clouds provide a global footprint, which enables resources and services to deploy in most major global regions. This enables data and processing activity to take place proximally to the region where the big data task is located. For example, if a bulk of data is stored in a certain region of a cloud provider, it's relatively simple to implement the resources and services for a big data project in that specific cloud region—rather than sustaining the cost of moving that data to another region.

1.5　Resilience

Data is the real value of big data projects, and the benefit of cloud resilience is in data storage reliability. Clouds replicate data as a matter of standard practice to maintain high availability in storage resources, and even more durable storage options are available in the cloud.

2.　Challenges to Big Data in a Cloud Environment

Cloud computing environments are typically provided as a service—Infrastructure as a Service (IaaS) offers virtualized computing resources, Platform as a Service (PaaS) offers application development resources, Software as a Service (SaaS) offers software, and Database as a Service (DBaaS) offers data resources. You can also find database-specific managed services, such as MySQL and Postgres as a Service.

When you sign up for any cloud computing service, you introduce a third-party entity into your business and digital ecosystem. When you delegate a task to a third-party service, you forfeit a certain level of control. In cloud computing, that means you're giving up control over many aspects of the digital environment, including your security perimeter.

Delegating responsibilities to cloud providers is a good choice for many professionals. However, there are some drawbacks, because you're essentially relying on your cloud provider to secure your cloud while ensuring compliance. Before choosing a cloud environment for your big data, make sure it meets the required compliance for the data you're processing and analyzing.

If you're analyzing transactional or ecommerce data that contains credit cardholder information, you'll need to comply with the Payment Card Industry Data Security Standard (PCI DSS), which regulates the security of cardholders' data. If you're processing healthcare information, you'll need to comply with the Health Insurance Portability and Accountability Act of 1996 (HIPAA), which protects personal health and medical information.

3. Choose the Right Cloud Deployment Model

So, which cloud model is ideal for a big data deployment? Organizations typically have four different cloud models to choose from: private, public, hybrid and multi-cloud (see Figure 10-2). It's important to understand the nature and tradeoffs of each model.

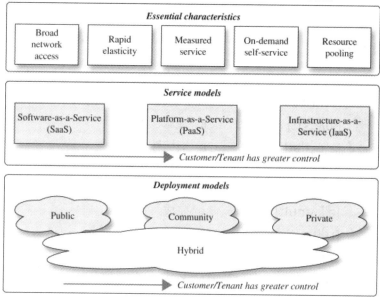

Figure 10-2   Cloud Model

3.1   Private cloud

Private clouds give businesses control over their cloud environment, often to accommodate specific regulatory, security or availability requirements. However, it is more costly because a business must own and operate the entire infrastructure. Thus, a private cloud might only be used for sensitive small-scale big data projects.

3.2   Public cloud

The combination of on-demand resources and scalability makes public cloud ideal for almost any size of big data deployment. However, public cloud users must manage the cloud resources and services it uses. In a shared responsibility model, the public cloud provider handles the security of the cloud, while users must configure and manage security in the cloud.

3.3   Hybrid cloud

A hybrid cloud is useful when sharing specific resources. For example, a hybrid cloud might enable big data storage in the local private cloud—effectively keeping data sets local and secure —and use the public cloud for compute resources and big data analytical services. However, hybrid clouds can be more complex to build and manage, and users must deal with all of the issues and concerns of both public and private clouds.

3.4   Multi-cloud

With multiple clouds, users can maintain availability and use cost benefits. However, resources and services are rarely identical between clouds, so multiple clouds are more complex to manage. This cloud model also has more risks of security oversights and compliance breaches than single public cloud use. Considering the scope of big data projects, the added complexity of multi-cloud deployments can add unnecessary challenges to the effort.

New Words

immediate	[ɪ'miːdɪət]	adj. 直接的；当前的
budget	['bʌdʒɪt]	n. 预算
agility	[ə'dʒɪləti]	n. 敏捷性，灵敏性
project	['prɒdʒekt]	n. 项目，工程；方案，计划
		v. 规划，计划
accomplish	[ə'kʌmplɪʃ]	v. 完成，达成
release	[rɪ'liːs]	v. 释放
enormous	[ɪ'nɔːməs]	adj. 巨大的，极大的
capital	['kæpɪtəl]	n. 资本
		adj. 资本的
expense	[ɪk'spens]	n. 开销，费用
footprint	['fʊtprɪnt]	n. 足迹；(计算机)占用的空间
region	['riːdʒən]	n. 地区
proximally	['prɒksɪməli]	adv. 最近地
resilience	[rɪ'zɪlɪəns]	n. 弹性
durable	['djʊərəbl]	adj. 耐用的；持久的
third-party	['θɜːd 'paːti]	adj. 第三方的
entity	['entəti]	n. 实体
forfeit	['fɔːfɪt]	v. 丧失，失去
responsibility	[rɪˌspɒnsə'bɪləti]	n. 责任，义务
essentially	[ɪ'senʃəli]	adv. 本质上，根本上
tradeoff	['treɪdˌɔːf]	n. 权衡
regulatory	['regjələtəri]	n. 监管
		adj. 监管的，具有监管权的
rarely	['reəli]	adv. 少有地，罕见地
identical	[aɪ'dentɪkl]	adj. 同一的，相同的
oversight	['əʊvəsaɪt]	n. 疏忽
effort	['efət]	n. 努力

Phrases

physical space	物理空间
software service	软件服务
big data project	大数据项目
rental model	租赁模式
pay-per-use model	按使用付费模型
a bulk of	许多
cloud provider	云提供商
virtualized computing resource	虚拟计算资源
third-party entity	第三方实体
digital ecosystem	数字生态系统
third-party service	第三方服务
give up	放弃
digital environment	数字环境
private cloud	私有云
public cloud	公共云
hybrid cloud	混合云
multiple cloud	多项云

Abbreviations

IaaS (Infrastructure as a Service)	基础设施即服务
DBaaS (Database as a Service)	数据库即服务
PCI (Payment Card Industry)	支付卡行业
DSS (Data Security Standard)	数据安全标准

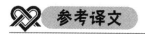 参考译文

云中的大数据

1. 云中大数据的优点

云为各种规模的企业带来了多种多样的重要好处。云中大数据的一些最直接和实质性

的好处包括以下几点。

1.1　可扩展性

典型的业务数据中心在物理空间、电力、冷却以及购买和部署构建大数据基础设施所需的大量硬件预算方面都面临着限制。相比之下，公共云管理着遍布全球数据中心的数十万台服务器。基础设施和软件服务已经有了，用户可以为几乎任何规模的大数据项目组装基础设施。

1.2　敏捷性

并非所有大数据项目都是一样的。一个项目可能需要 100 台服务器，而另一个项目可能需要 2000 台服务器。使用云，用户就可以按需使用资源来完成任务，然后在任务完成时释放这些资源。

1.3　成本

商业数据中心是一项巨大的资本支出。除了硬件，企业还必须支付设施、电力、日常维护等费用。云将所有这些成本转换为灵活的租赁模式，在该模式中，资源和服务可按需提供，并遵循按使用付费的模式。

1.4　可访问性

许多云提供全球可访问性，这使资源和服务能够部署在全球大多数主要地区。这使得数据和处理活动能够在大数据任务所在的区域附近进行。例如，如果大量数据存储在云提供商的某个区域，那么在该特定云区域中实施大数据项目的资源和服务相对简单——而不需要承担将数据移动到另一个区域的成本。

1.5　弹性

数据是大数据项目真正的价值所在，云弹性的好处在于数据存储的可靠性。云复制数据是一种标准做法来保持存储资源的高可用性，并且在云中提供更持久的存储选项。

2. 云环境下大数据面临的挑战

云计算环境通常以服务的形式提供——基础设施即服务(IaaS)提供虚拟化计算资源，平台即服务(PaaS)提供应用程序开发资源，软件即服务(SaaS)提供软件，数据库即服务(DBaaS)提供数据资源。你还可以找到特定数据库的托管服务，例如 MySQL 和 Postgres即服务。

当注册任何云计算服务时，就会将第三方实体引入到你的业务和数字生态系统中。当将任务委派给第三方服务时，你将失去一定程度的控制权。在云计算中，这意味着你将放弃对数字环境许多方面的控制，包括你的安全边界。

对许多专业人士来说，将责任交给云提供商是很好的选择。但是，这样做也存在一些缺点，因为你基本上依赖于云提供商来保护云，同时确保合规性。在为你的大数据选择云环境之前，请确保它满足你正在处理和分析的数据所需的合规性。

如果你在分析包含信用卡持卡人信息的交易或电子商务数据，则需要遵守支付卡行业数据安全标准(PCI DSS)，该标准规定了持卡人数据的安全性。如果你正在处理医疗保健信息，你需要遵守 1996 年的健康保险流通与责任法案 (HIPAA)，该法案保护个人健康和医疗信息。

3. 选择合适的云部署模式

哪种云模型最适合大数据部署？组织通常有四种不同的云模型可供选择：私有云、公共云、混合云和多项云。了解每个模型的性质并权衡利弊很重要(见图10-2)。

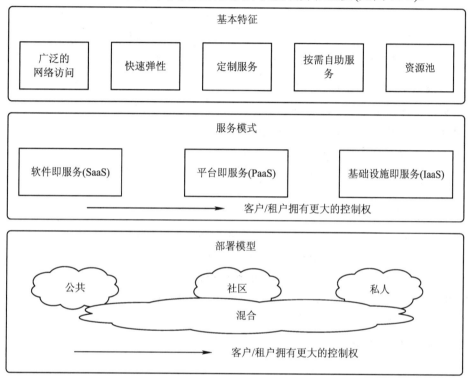

图 10-2　云模型

3.1　私有云

私有云让企业能够控制他们的云环境，这通常是为了满足特定的监管、安全或可用性要求。然而，私有云的成本更高，因为企业必须拥有和运营整个基础设施。因此，私有云可能仅用于敏感的小规模大数据项目。

3.2　公共云

按需使用资源和可扩展性的结合使公共云成为几乎任何规模的大数据部署的理想选择。但是，公共云用户必须管理其使用的云资源和服务。在责任共担模型中，公共云提供商处理云的安全性，而用户必须配置和管理云中的安全性。

3.3　混合云

在共享特定资源时，混合云很有用。例如，混合云可能会在本地私有云中实现大数据存储——有效地保持数据集的本地性和安全性——并使用公共云提供计算资源和大数据分析服务。然而，混合云的构建和管理更加复杂，用户必须处理公共云和私有云的所有问题和担忧。

3.4 多项云

使用多项云，用户可以保持可用性并使用成本优势。但是，云之间的资源和服务很少相同，因此多项云的管理更加复杂。与单一的公共云使用相比，这种云模型还具有更多的安全疏忽和违规风险。考虑到大数据项目的范围，多项云部署增加的复杂性会给工作带来不必要的挑战。

Bigdata **Exercises**

〖Ex. 1〗 根据 Text A 回答以下问题。

1. Why were only large-scale businesses able to utilize big data during its early days?
2. What is it becoming quite common? Why?
3. What can AI programs do now?
4. What do businesses use the necessary analytics tools to do?
5. What is called dark data?
6. What are chief data officers expected to do?
7. If we can crunch billions of data at once in just a few minutes, what can we do?
8. Why is there no doubt open source software is cheaper?
9. Why does edge computing delivers a better performance?
10. What are chatbots now being deployed by companies to do?

〖Ex. 2〗 根据 Text B 回答以下问题。

1. What does a typical business data center face?
2. What can users do with cloud?
3. What is data? What is the benefit of cloud resilience
4. What is a good choice for many professionals? Why are there some drawbacks?
5. If you're processing healthcare information, what will you need to do?
6. What are the four different cloud models organizations typically have to choose from?
7. Why is a private cloud more costly?
8. What makes public cloud ideal for almost any size of big data deployment?
9. What are the disadvantages of hybrid clouds?
10. What can users do with multiple clouds?

〖Ex. 3〗 把下列词组翻译成中文。

1. cloud technology _____ 1. _____
2. computing cost _____ 2. _____
3. customer experience _____ 3. _____

4. data flow   4. _____

5. edge computing   5. _____

6. lag time   6. _____

7. log data   7. _____

8. open source software   8. _____

9. big data project   9. _____

10. cloud provider   10. _____

〖Ex. 4〗 把下列单词或词组翻译成英文。

1. 数字生态系统   1. _____

2. 私有云   2. _____

3. 公共云   3. _____

4. n. 机器人   4. _____

5. v. 实现，达到   5. _____

6. n. 机密性   6. _____

7. n. 互联网   7. _____

8. adj. 预建的，预制的   8. _____

9. n. 敏捷性，灵敏性   9. _____

10. n. 监管   adj. 监管的   10. _____

〖Ex. 5〗 翻译句子。

1. Virtualization is often closely linked with cloud technology.

2. Cloud computing technology provides a new perspective for data mining.

3. Hadoop cloud storage technology can solve this problem.

4. This is why quantum computing is so powerful.

5. It is important to find a right balance between operational efficiency, data security, infrastructure cost and business continuity.

6. Infrastructure as a service offers computing capabilities and basic storage as standardized services over the network.

7. Hybrid clouds combine both public and private cloud models.

8. Management responsibilities are shared by the public cloud provider and the business itself.

9. Facebook has their own infrastructure rather than using a third party cloud provider.

10. Cloud computing promises a number of advantages for the deployment of software applications.

Bigdata **Reading Material**

Skills Required to Become a Big Data Developer

During the past few years, we are getting more and more familiar with[①] the term big data. We have seen how big data becomes the king in the IT world. Big data will continue to rule the world even after the next few decades.

1. What Is a Big Data Developer?

A big data developer is the one who is responsible for developing Hadoop applications. It typically serves the big data needs of an organization he is working in and works to solve the big data problems and requirements.

A big data developer must be skilled enough to manage the complete Hadoop solution life cycle[②], including platform selection, designing technical architecture, requirement analysis, application development and design, testing, and deployment. He is responsible for the actual coding of Hadoop applications.

2. Roles and Responsibilities of Big Data Developer

The roles and responsibilities of the big data developer who is responsible for programming Hadoop applications in the big data domain are:

- Load the data from disparate data sets.
- High-speed querying.
- Propose best practices and standards.
- Design, build, install[③], configure and support Hadoop.
- Maintain security and data privacy.
- Manage and deploy[④] HBase.
- Perform an analysis of the vast number of data stores and uncover insights.
- Be responsible for Hadoop development and implementation.
- Be responsible for creating scalable and high-performance[⑤] web services for tracking data.
- Translate complex technical and functional requirements into detailed designs.

---

① familiar with：熟悉
② life cycle：生命周期
③ install [ɪnˈstɔːl]　vt. 安装
④ deploy [dɪˈplɔɪ]　v. 部署
⑤ high-performance: [hai-pəˈfɔːməns]　adj. 高效能，高性能的

• Propose design changes and suggestions① to various processes and products.

3. Skills Required to Become Big Data Developer

3.1 Hadoop-based technologies

The rise of big data in the early 21th century gave birth to a new framework called Hadoop. All the credit goes to Doug Cutting for introducing a framework that stores and processes data in a distributed manner and performs parallel processing.

Hadoop prevails to be the foundation② of other rising big data technologies. Learning Hadoop is the first step towards becoming a successful big data developer. Hadoop is not a single term, instead, it is a complete ecosystem. The Hadoop ecosystem contains a number of tools that serve different purposes.

For boosting your career as a big data developer, mastering some big data tools are a must.

Big data tools which you need to master are:

• HDFS (Hadoop Distributed File System): HDFS is the storage layer in Hadoop. It stores data across a cluster of commodity hardware. Before learning Hadoop, one should have knowledge of Hadoop HDFS as it is one of the core components of the Hadoop framework.

• YARN: YARN is responsible for managing resources amongst applications running in the Hadoop cluster. It performs resource allocation and job scheduling in the Hadoop cluster. The introduction of YARN makes Hadoop more flexible, efficient and scalable.

• MapReduce: MapReduce is the heart of the Hadoop framework. It is a parallel processing framework that allows data to be processed in parallel across clusters of inexpensive hardware.

• Hive: Hive is an open-source data warehousing tool built on top of Hadoop. With Hive, developers can perform queries on the vast amount of data stored in Hadoop HDFS.

• Pig: It is a high-level scripting language③ used for data transformation on the top of Hadoop. It is used by researchers for programming.

• Flume: Flume is a reliable, distributed tool for importing large amounts of streaming data such as events, log data, etc from different web servers to the Hadoop HDFS.

• Sqoop: Sqoop is a big data tool used for importing and exporting data④ from relational databases such as MySQL, Oracle, etc to Hadoop HDFS or vice versa⑤.

• ZooKeeper: It is a distributed coordination⑥ service that acts as a coordinator among the distributed services running in the Hadoop cluster. It is responsible for managing and

---

① suggestion [sə'dʒestʃən] n. 建议

② foundation [faʊn'deɪʃn] n. 基础

③ scripting language：脚本语言

④ importing and exporting data：导入和导出数据

⑤ vice versa：反之亦然

⑥ coordination [kəʊˌɔ:dɪ'neɪʃn] n. 协调(能力)；配合

coordinating a large cluster of machines.

• Oozie: Oozie is a workflow scheduler[①] for managing Hadoop jobs. It binds[②] multiple jobs into a single unit of work and helps in accomplishing a complete task.

3.2 Apache Spark

Real-time processing[③] with rapid action is the need of the world. Whether it is a fraud detection system or recommendation system, every one of them requires real-time processing. For a big data developer, it is very important to be familiar with the real-time processing framework.

Apache Spark is a real-time distributed processing framework with in-memory computing capabilities. So Spark is the best choice for big data developers to be skilled in any of the one real-time processing frameworks.

3.3 SQL

SQL (Structure Query Language) is the data-centered language used to structure, manage and process the structured data stored in databases.

Since SQL is the base of the big data era, so the knowledge of SQL is an added advantage to the programmers while working on big data technologies. PL/SQL[④] is also widely used in the industry.

3.4 NoSQL

The organizations are generating data at rapid speeds. The amount of data has grown beyond our imagination[⑤]. The requirements of the organizations are now extended from structured to unstructured data.

To meet the increasing requirements of the organizations, NoSQL databases were introduced. The NoSQL database can store and manage large amounts of structures, semi-structured, and unstructured data.

Some of the prominently[⑥] used NoSQL databases are:

• Cassandra: Cassandra is a NoSQL database that provides scalability and high availability without compromising performance. It is a perfect platform for mission-critical data. Cassandra provides fast and random read/writes. It provides Availability and Partitioning out of CAP[⑦].

• HBase: HBase is a column-oriented NoSQL database built on top of Hadoop HDFS. It

---

① workflow scheduler：工作流调度程序

② bind [baɪnd] v. 绑定，捆绑

③ real-time processing：实时处理

④ PL/SQL：Procedural Language/SQL，即过程化 SQL 语言

⑤ imagination [ɪˌmædʒɪˈneɪʃn] n. 想象(力)

⑥ prominently [ˈprɒmɪnəntli] adv. 显著地，重要地

⑦ CAP: Consistency, Availability, Partition-tolerance，即一致性、可用性、分区容错性

provides quick random[①] real-time read or writes access to the data stored in the Hadoop File System. HBase provides Consistency[②] and Partitioning out of CAP.

• MongoDB: MongoDB is a general-purpose document-oriented NoSQL database. It is a NoSQL database that provides high availability, scalability, and high performance. MongoDB provides Consistency and Partitioning out of CAP.

A professional with knowledge of NoSQL databases will never go out of fashion.

3.5　Programming language

For being a big data developer, you must have good hands in coding. You must have knowledge of data structures, algorithms, and at least one programming language.

There are various programming languages like Java, R, Python, Scala, etc. that caters to the same purposes. All the programming languages have different syntax but the logic remains the same.

For a beginner, I suggest that you should go with Python as it is simple to learn and it is a statistical language.

3.6　Data visualization tools

The big data professionals must have the ability to interpret the data by visualizing it. This requires mathematic and science edge to easily understand complex large data with creativity[③] and imagination.

There are some prominent[④] data visualization tools like QlikView, Tableau, QlikSense that help in understanding the analysis performed by various analytics tools. Learning the visualization tools adds an advantage for you if you want to boost your data analytics and visualization skills.

3.7　Data mining

Data mining is very important when we talk about extracting, storing, and processing vast amounts of data. For working with big data technologies, you must be familiar with data mining tools like Apache Mahout, Rapid Miner, KNIME, etc.

3.8　Machine learning

Machine learning is the hottest field of big data. ML helps in developing a recommendation, personalization[⑤], and classification systems. With the advancements in technologies, professionals with machine learning skills for predictive and prescriptive analysis are scarce.

For being a successful data analyst one should have a good hand in the machine learning algorithms.

---

① random ['rændəm]　adj. 随机的；任意的
② consistency [kən'sɪstənsi]　n. 一致性；连贯性
③ creativity [ˌkriːeɪ'tɪvəti]　n. 创造性，创造力
④ prominent ['prɒmɪnənt]　adj. 突出的，杰出的
⑤ personalization ['pɜːsənəlaɪzeɪʃn]　n. 个性化

3.9 Statistical and quantitative analysis

Big data is all about numeric digits. Quantitative and statistical analysis is the most important part of big data analysis.

The knowledge in statistics and mathematics helps in understanding core concepts such as probability distribution[1], summary statistics, random variables[2]. The knowledge of different tools like R, SAS, SPSS, etc. makes you different from others standing in the queue.

3.10 Linux or Unix or Solaris or MS-Windows

Various operating systems[3] are used by a wide range of industries. Unix and Linux are the most used operating systems. The big data developer needs to master at least one of them.

3.11 Creativity and problem solving

You must have the problem-solving ability and creative mind[4] while working in any domain. The implementation of different big data techniques for efficient solutions needed both of these qualities in the professionals.

3.12 Business knowledge

For working in a domain, one of the most important skills is the knowledge of the domain he is working in. To analyze any data or develop any application, one should have the business knowledge to make analyses or development profitable.

---

① probability distribution：概率分布

② random variable：随机变量

③ operating system：操作系统

④ creative mind：创造性思维